AF328856

Springer Tracts in Modern Physics

Volume 287

Series Editors

Mishkatul Bhattacharya, Rochester Institute of Technology, Rochester, NY, USA

Yan Chen, Department of Physics, Fudan University, Shanghai, China

Atsushi Fujimori, Department of Physics, University of Tokyo, Tokyo, Japan

Mathias Getzlaff, Institute of Applied Physics, University of Düsseldorf, Düsseldorf, Nordrhein-Westfalen, Germany

Thomas Mannel, Emmy Noether Campus, Universität Siegen, Siegen, Nordrhein-Westfalen, Germany

Eduardo Mucciolo, Department of Physics, University of Central Florida, Orlando, FL, USA

William C. Stwalley, Department of Physics, University of Connecticut, Storrs, USA

Jianke Yang, Department of Mathematics and Statistics, University of Vermont, Burlington, VT, USA

Springer Tracts in Modern Physics provides comprehensive and critical reviews of topics of current interest in physics. The following fields are emphasized:

– Particle and Nuclear Physics
– Condensed Matter Physics
– Light Matter Interaction
– Atomic and Molecular Physics

Suitable reviews of other fields can also be accepted. The Editors encourage prospective authors to correspond with them in advance of submitting a manuscript. For reviews of topics belonging to the above mentioned fields, they should address the responsible Editor as listed in "Contact the Editors".

Carlos Torres-Torres • Geselle García-Beltrán

Optical Nonlinearities in Nanostructured Systems

 Springer

Carlos Torres-Torres
Instituto Politécnico Nacional
ESIME Zacatenco
Mexico City, Mexico

Geselle García-Beltrán
Instituto Politécnico Nacional
ESIME Zacatenco
Mexico City, Mexico

ISSN 0081-3869 ISSN 1615-0430 (electronic)
Springer Tracts in Modern Physics
ISBN 978-3-031-10823-5 ISBN 978-3-031-10824-2 (eBook)
https://doi.org/10.1007/978-3-031-10824-2

© The Editor(s) (if applicable) and The Author(s), under exclusive license to Springer Nature Switzerland AG 2022
This work is subject to copyright. All rights are solely and exclusively licensed by the Publisher, whether the whole or part of the material is concerned, specifically the rights of translation, reprinting, reuse of illustrations, recitation, broadcasting, reproduction on microfilms or in any other physical way, and transmission or information storage and retrieval, electronic adaptation, computer software, or by similar or dissimilar methodology now known or hereafter developed.
The use of general descriptive names, registered names, trademarks, service marks, etc. in this publication does not imply, even in the absence of a specific statement, that such names are exempt from the relevant protective laws and regulations and therefore free for general use.
The publisher, the authors, and the editors are safe to assume that the advice and information in this book are believed to be true and accurate at the date of publication. Neither the publisher nor the authors or the editors give a warranty, expressed or implied, with respect to the material contained herein or for any errors or omissions that may have been made. The publisher remains neutral with regard to jurisdictional claims in published maps and institutional affiliations.

This Springer imprint is published by the registered company Springer Nature Switzerland AG
The registered company address is: Gewerbestrasse 11, 6330 Cham, Switzerland

Foreword

Interest in the field of nonlinear optics has grown exponentially in recent years, as there is a growing number of scientists who participate in the study of new non-linear phenomena and in the development of new nonlinear devices in the field of photonics. In addition to being a very extensive field that includes the fundamentals of the interactions of light with matter, particular processes such as quantum confinement and surface plasmon resonance tools can be considered as an important contribution to the tuning and amplification of nonlinear characteristics. This field of study has grown in a very large proportion, so it would be impossible to cover all the fields of study of nonlinear optical effects in a single book because nonlinear optics also turns out to be complex. That is why this book is mainly focused on the study of nonlinear optics with the intention of providing a broad overview of the nonlinearities of various attractive nanostructures. It includes topics such as the effects represented by the interaction of multi-wave mixing, description of the various methods for measuring second- and third-order nonlinearities, properties and description of the optical parameters associated with plasmonic nanostructures, low-dimensional carbon-based materials, and representative semiconductors. These topics have been selected in order to trace a route for the design of important applications related to nonlinear optics. This book emphasizes the fundamental concepts behind second- and third-order nonlinear optical effects including the size and shape of particular nanostructures, as well as the potential applications of nanomaterials. So, this text is intentionally written to provide an introduction and general background to basic ideas for people who specialize in the discipline of nonlinear optical effects, as well as those seeking to become familiar with it.

Pennsylvania State University
State College, PA, USA

Dr. Nestor Perea-Lopez

Preface

Since multiphonic nonlinearities cover a very large and diverse area in the study of nonlinear optics, the present book has been written with the purpose of pointing out second- and third-order nonlinear optical effects in various nanostructured systems. This book is focused on the description of the properties of the systems from the non-linear interactions associated with the mixing of optical waves. Thus, detailed existing information from the field of nonlinear optics and nanoscale processes have been compiled and analyzed. For this, a review of the theoretical descriptions of second- and third-order optical nonlinearities in specific nanostructured systems is presented and their respective susceptibility tensor in the presence of mixed frequencies of optical waves is discussed. And their respective susceptibility tensor in the presence of mixed frequencies of optical waves has been discussed. A large number of factors that influence the real and imaginary parts of the nonlinear susceptibility have been identified. Furthermore, a route is established on how, through the interactions derived from the mixing of optical waves, relevant information can be obtained for the characterization of nanostructures and the design of new applications in various technological fields. In addition to the study of ultrafast phenomena related to specific nanostructured systems, the complexity of the localized surface plasmon resonance phenomenon characteristic of metallic nanostructured systems is indicated, which is also related to the size of the nanostructures. Moreover, quantum confinement effects derived from semiconductor nanostructured systems and the allotropic property of carbon are studied.

In the present work, the fundamental physics related to electromagnetic theory, absorption spectra, and quantum energy level diagrams involved in nonlinear optics have been analyzed to be a base for future research.

Finally, the authors kindly acknowledge the financial support of the Instituto Politécnico Nacional (IPN, Mexico) and the Consejo Nacional de Ciencia y Tecnología (CONACyT, Mexico).

Instituto Politécnico Nacional   Carlos Torres-Torres
ESIME Zacatenco   Geselle García-Beltrán
Mexico City, Mexico

Contents

Abbreviations

AlAs	Aluminum arsenide
AlSb	Aluminum antimonide
BaO	Barium oxide
Cd_3As_2	Cadmium arsenide
Cd_3P_2	Cadmium phosphide
CdS	Cadmium sulfide
CdSe	Cadmium selenide
CdTe	Cadmium telluride
CeO_2	Cerium oxide
CuInSe	Copper indium selenide
CuO	Cupric oxide
DFG	Difference-frequency generation
GaAs	Gallium arsenide
GaN	Gallium nitride
GaP	Gallium phosphide
GaSb	Gallium antimonide
InAs	Indium arsenide
InP	Indium phosphide
InSb	Indium antimonide
MoS_2	Molybdenum disulfide
NiO	Nickel oxide
OPA	Optical parametric amplification
PbS	Lead sulfide
PbSe	Lead selenide
PbTe	Lead telluride
SFG	Sum-frequency generation
SiC	Silicon carbide
SiGe	Silicon germanium
SiO_2	Silicon dioxide

$SrTiO_3$	Stronium titanate
TiO_2	Titanium dioxide
ZnO	Zinc oxide
ZnS	Zinc sulfide
ZnSe	Zinc selenide
ZnTe	Zinc telluride

List of Figures

Chapter 1
Methods for Measuring Nonlinear Optical Properties

1.1 Introduction

Nonlinear optics is the branch in charge of studying the phenomena that involve the interaction between intense electromagnetic radiation with a material. Optical nonlinearities are defined because the incident field results in the redistribution of charge from the material. The rearrangement of the charge distribution gives rise to changes in the optical properties of the media. The description of nonlinear optical phenomena is defined through the polarization of the medium. The effect produced by the third-order nonlinear susceptibility polarization determines a contribution to the change in refractive index in a transparent dielectric medium. The optical nonlinearities of the materials are analyzed through laser systems since it is the best option to produce an energy source with sufficient intensity to generate nonlinear optical phenomena.

Interest in nonlinear optical materials is on a continuous increase. Through this, linear optical effects and higher-order effects treated at higher energy levels known as nonlinear optical effects are analyzed. Nonlinear optical effects are derived because of electronic interactions within the studied matter. The interactions generate higher-order optical harmonics, which are responsible for producing nonlinear optical effects. Nonlinear optical processes have focused on the understanding of various materials such as organic, inorganic, polymeric, biological, and nanostructured. The nonlinear optical responses are governed under the conditions of symmetry of the materials. Such conditions of symmetry demonstrate that second-order nonlinear optical effects appear in materials with molecules that lack a center of symmetry. On the other hand, materials without symmetric requirements exhibit third-order nonlinear optical effects. Thus, it is possible to use a wide variety of materials for third-order nonlinear optics. Recently, photoinduced tests have been exploited more and more due to the wide variety of optoelectronics and photonic applications that can be developed due to the optical nonlinearities of materials. For

© The Author(s), under exclusive license to Springer Nature Switzerland AG 2022

C. Torres-Torres, G. García-Beltrán, *Optical Nonlinearities in Nanostructured Systems*,
Springer Tracts in Modern Physics 287, https://doi.org/10.1007/978-3-031-10824-2_1

this reason, knowledge of the fundamental and material aspects of third-order nonlinear optical effects is of utmost importance. Furthermore, the study of the nonlinear optics of materials has been shown to involve multiple disciplines. In addition, it is important to study the theoretical parameters and physical processes that determine third-order nonlinear optical responses in materials at the macro- and nanoscale.

Third-order optical nonlinearities are governed by several factors, such as third-order nonlinear optical susceptibilities determined from various methods. Such methods include third harmonic generation, optical wave coupling, Kerr optical gate, and self-diffraction techniques. These techniques have quite unfamiliar characteristics from each other. These characteristics are highly dependent on the experimental conditions used. These conditions are incident wavelength, temporal pulse conditions, energy of the laser irradiance, conditions of the environment surrounding the sample, and state of matter in which the material is found.

The resonance of third-order optical nonlinearities is determined across wavelength. This is now facilitated because there is a wide range of laser systems that have a selection of wavelengths for the measurement of third-order nonlinear optical susceptibility of materials. The extensive selection of measurement wavelengths enables the determination of resonant and nonresonant contributions to third-order optical nonlinearities exhibited by a material. Nonlinear third-order resonant optical susceptibility is derived from the absorption of nonlinear medium near the incident wavelengths. The resonance presented by the materials is associated with the increase in the nonlinear third-order optical susceptibility values. Consequently, through resonance effects and their absence, the intensity of third-order nonlinear optical susceptibility can fluctuate by several orders of magnitude. One of the relevant reasons for the study of third-order nonlinear optical phenomena is the nonlinear refractive index and absorption coefficient dependent on the irradiance of the materials. The third-order nonlinear optical coefficients are considered of importance due to the various fields of application. Some examples of nonlinear applications are optical switches, optical limiting, optical communications, logic gates, and quantum communication.

In summary, it has been determined through the available literature that third-order nonlinear optical materials can be used for controlling light by light. However, there are also limitations related to the interacting chemical, physical, and optical mechanisms in the case of organic material changes of any given material system. So, semiconductor structures such as semiconductor optical amplifiers and passive quantum well devices provide a breakthrough in the field of materials science. Finally, the need for nonlinear organic optical materials to demonstrate developing applications that involve superior compatibility with integrated hybrid optical structures is highlighted.

Interest in the field of optical nonlinearities has grown steadily since its inception, spanning fundamental light–matter interactions, to applications that include frequency conversions through laser systems. The field of nonlinear optics is very extensive and begins from the moment some main wave properties of light are modified due to the interaction with the material and the electromagnetic wave.

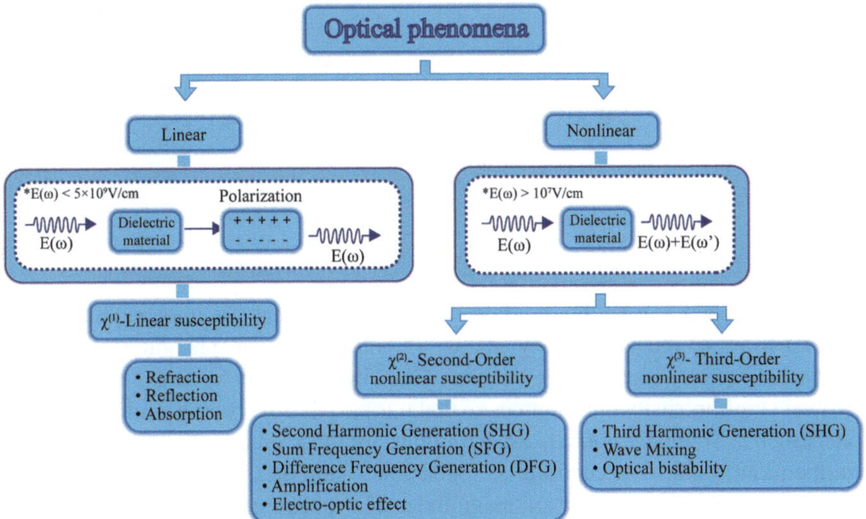

Fig. 1.1 Classification of optical phenomena according to the order of susceptibility

Electromagnetic waves in a medium interact through nonlinear polarization. In general, a nonlinear optical effect originates from the interaction of the propagation of light in a medium with nonlinear susceptibilities as coupling coefficients. When a beam of light irradiates a crystalline material, its associated electric field produces a polarization vector in the crystal in phase with the incident signal. If the material has nonlinear characteristics, the vector associated with electrical polarization is no longer sinusoidal. The signal is transmitted through the material and then becomes a compound function [1]. Moreover, some cases include interactions of waves other than electromagnetic, such as acoustic and mechanical. In contrast, for linear optics, in which the interaction of light with matter does not modify the wave properties, linear effects occur when the intensity of the light is moderate or low. For nonlinear optical effects, the higher the order of the nonlinear process to be generated, the higher the intensity of the incident beam is necessary. Thus, it is possible to generate various optical phenomena depending on the material and magnitude of the susceptibility.

As a harmonic wave passes through a nonlinear medium, wave polarization occurs, causing the material to respond to the applied electric field with charge redistribution. If the effect were only linear, the electrical polarization wave would be represented through an oscillatory field together with an incident light. Moreover, the irradiated light would be represented as a refracted wave propagating at a different speed regarding the external medium, conserving the frequency of the incident light. As shown in Fig. 1.1, this in turn expresses the type of phenomenon associated with the order of optical susceptibility.

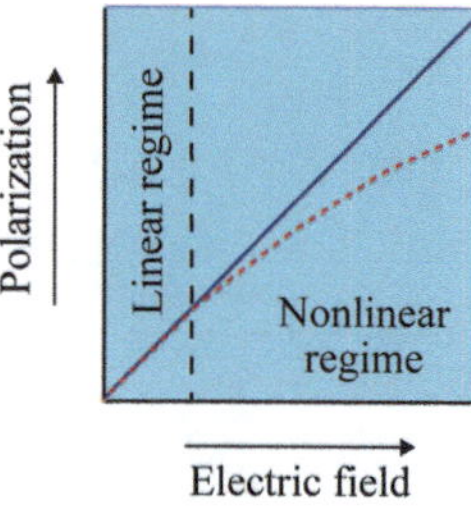

Fig. 1.2 Polarization response due to the change in magnitude of the induced electric field

It is important to note that nonlinear optical phenomena are imperceptible when the radiation source is not very intense. Nonlinear optical phenomena manifest themselves at very high light intensities. Thus, through laser systems, it is possible to obtain high-intensity light sources so that it is possible to observe nonlinear optical phenomena. Nonlinear optical susceptibilities are those greater than the first- order. Furthermore, the susceptibilities generate nonlinear optical phenomena corresponding to their order of magnitude. The phenomena are associated with quantities dependent on the square or higher orders dependent on the intensity of the electric field. In Fig. 1.2, the polarization response corresponding to the magnitude increase in the induced electric field can be seen.

In addition, nonlinear optical phenomena have unique characteristics of their kind. These characteristics establish the conversion of optical frequency and the modification of the properties of the material through its dependence on the field.

1.2 Experimental Techniques in Second-Order Optical Nonlinearities Evaluations

Second-order nonlinear optical effect-generating materials are generally non-centro-symmetric crystals. Thus, for a crystalline structure to show a second-order nonlinear effect, it must have a polar axis. This is associated with the absence of an axis of symmetry in the crystal.

There are some classes of crystals that meet the non-centrosymmetric condition. However, in many of the crystals the effect is too small to be measured experimentally. Equally are characterized dipolar molecules that exhibit second-order nonlinear optical effects by having the characteristic that they are polarizable, which establishes that they have high electron mobility; in addition, they must have good chemical stability.

Nonlinear optics describes the behavior of light in media where the dielectric component of polarization responds to the nonlinear shape of the electric field of light. Furthermore, in the nonlinear regime, it is possible to expand the polarization in a Taylor series, in terms of the applied electric field. So, the induced polarization can be written as

$$P_{\text{NL}} = \left[\chi^{(1)} E + \chi^{(2)} E^2 + \ldots \right] \tag{1.1}$$

where $\chi^{(1)}$ is the optical susceptibility that derives linear optical effects, and $\chi^{(2)}$ refers to the nonlinear second-order optical susceptibility that derives nonlinear phenomena.

On the other hand, the optical susceptibilities expressed in the frequency domain with complex sets are associated such that the nonlinear second-order susceptibilities, in the frequency domain, to define their real and imaginary part, can be expressed as

$$\chi^{(2)} = \chi^{(2)'} + i\,\chi^{(2)''} \tag{1.2}$$

In this case, the intrinsic resonances of a material with the imaginary part of the second-order nonlinear susceptibility are associated. Similarly, the transitions of energy levels of the material are also related to resonances. Furthermore, the imaginary part of the second-order nonlinear optical susceptibility reaches a high peak near the resonant frequencies, while the real part shows dispersive behavior.

The real frequency-dependent optical susceptibility gives way to so-called parametric optical interactions, while the optical interactions associated with the imaginary part of a complex frequency-dependent susceptibility are termed nonparametric. In nonparametric interactions, the state of the material change is the same as the optical energy change. The above results in a process connected to resonant transitions in the material. On the other hand, when the interaction is only dependent on the real frequency, the state of the material in the medium remains unchanged at the same as the total optical energy. So, the process does not interfere with the energy between the optical field and the medium. With this, it is possible to explain a parametric linear process, such that it is possible to conserve energy through an optical frequency component. Similarly, the linear susceptibility $\chi^{(1)}$ is a function of a single frequency, so it is not possible to couple fields of different frequencies. It is observed that the susceptibility $\chi^{(2)}$ corresponds to nonlinear second-order phenomena. Moreover, $P_{\text{NL}}^{(2)}$ corresponds to the quadratic dependence of the electric field, which can be expressed as

$$P_{\text{NL}}^{(2)} = \chi^{(2)} \text{EE} \tag{1.3}$$

Similarly, the equation for the electric field is expressed as

$$E(r,t) = E_0 \left(e^{i(kr - wt)} + c \right) \tag{1.4}$$

The most common process in nonlinear optics is the second-order process such that substituting (1.4) in the quadratic nonlinear polarization equation, and also developing the result of the substitution, the following pair of equations is obtained:

$$P_{\mathrm{NL}}^{(2)} = \chi^{(2)} E_0 \left(e^{i(kr-wt)} + c \right) E_0 \left(e^{i(kr-wt)} + c \right) \tag{1.5}$$

$$P_{\mathrm{NL}}^{(2)} = 2\chi^{(2)} E_0^2 + \chi^{(2)} E_0^2 \left(e^{i(kr-wt)} + c \right) \tag{1.6}$$

Furthermore, if we use two electric fields at different frequencies $E_1(r, t)$ and $E_2(r, t)$, where

$$E_1(r,t) = E_{0,1} \left(e^{i(k_1 r - w_1 t)} + c \right) \tag{1.7}$$

$$E_2(r,t) = E_{0,2} \left(e^{i(k_2 r - w_2 t)} + c \right) \tag{1.8}$$

Where the total incident field is $E(r, t) = E_1(r, t) + E_2(r, t)$, like this

$$P_{\mathrm{NL}}^{(2)} = 2\chi^{(2)} E_{0,1}^2 + 2\chi^{(2)} E_{0,2}^2 + \chi^{(2)} E_{0,1}^2 \left(e^{2i(k1r - w1t)} + c \right)$$

$$+ \chi^{(2)} E_{0,2}^2 \left(e^{2i(k2r - w2t)} + c \right)$$

$$+ 2\chi^{(2)} E_{0,1} E_{0,2} \left(e^{i(k_1 + k_2)r - (w_1 + w_2)t} + c \right)$$

$$+ 2\chi^{(2)} E_{0,1} E_{0,2} \left(e^{i(k_1 - k_2)r - (w_1 - w_2)t} + c \right) \tag{1.9}$$

On the other hand, the energy exchange between the frequency components that results from a nonlinear parametric process, which in turn results in the nonlinear coupling between the components, is produced. Furthermore, the sum of the energies of all the interacting frequency components is conserved. Then, in a parametric process, there can be an energy change in the individual frequency components. However, relating to the frequency components, the total optical energy contained is conserved because there is no energy exchange between the optical field and the middle.

Then, if Eq. (1.6) is compared with (1.8), it is possible to observe the contributions associated with the second harmonic generation and the optical rectification process. Similarly, the terms that represent phenomena associated with the operation of frequencies are observed. So, the polarization described in Eq. (1.9) can be used to describe the second-order nonlinear optical processes such as second harmonic generation, sum-frequency generation, difference-frequency generation, optical parametric amplification, optical parametric generation, optical rectification, and Pockels effect.

In the sum-frequency generation phenomenon, two input beams travel through the nonlinear medium, such that the energies of the beams add up to create a new beam with higher energy, as shown in Fig. 1.3.

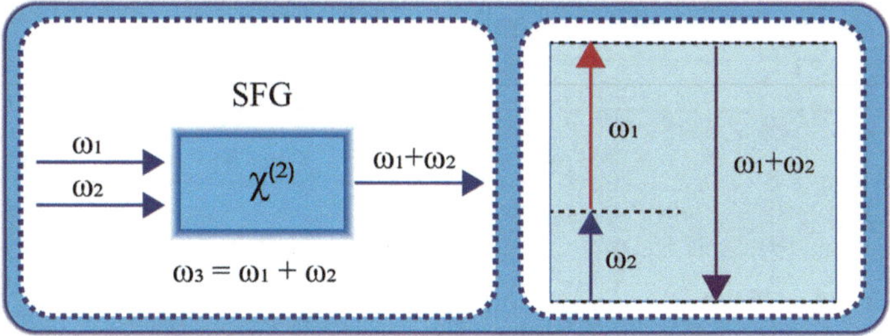

Fig. 1.3 Schematic diagram of the sum-frequency generation (SFG) and energy-level representation of the SFG

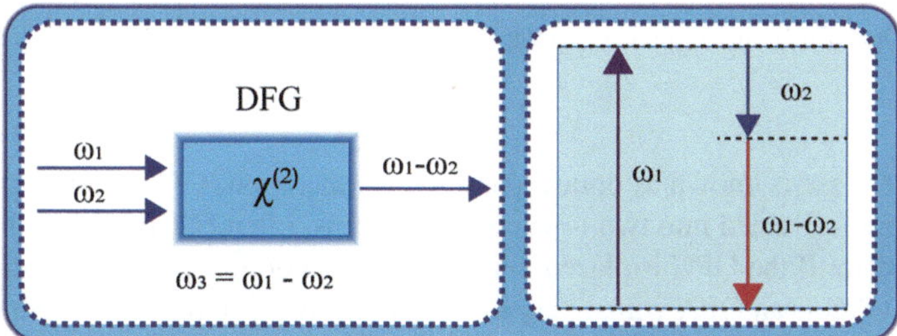

Fig. 1.4 Schematic diagram of the difference-frequency generation (DFG) and energy-level representation of the DFG

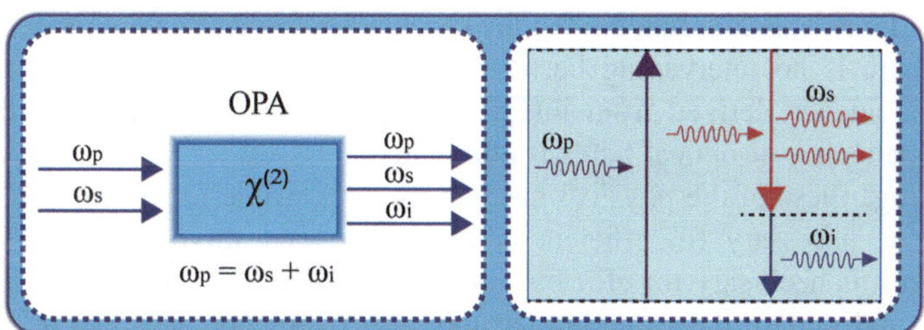

Fig. 1.5 Optical parametric amplification (OPA) and energy-level representation of the OPA

On the other hand, for the generation of the difference-frequency generation (DFG) phenomenon, a beam with lower energy is created by subtracting the lower energy from the higher energy, as shown in Fig. 1.4 [2].

In addition, the phenomenon known as optical parametric amplification is a second-order nonlinear optical process that establishes the transfer of energy between beams at different frequencies. The phenomenon consists of the transfer of energy from a high-intensity beam to a lower-intensity beam, which results in an amplification of the signal. Due to the interaction of the beams, a third light beam is generated [3], as shown in Fig. 1.5.

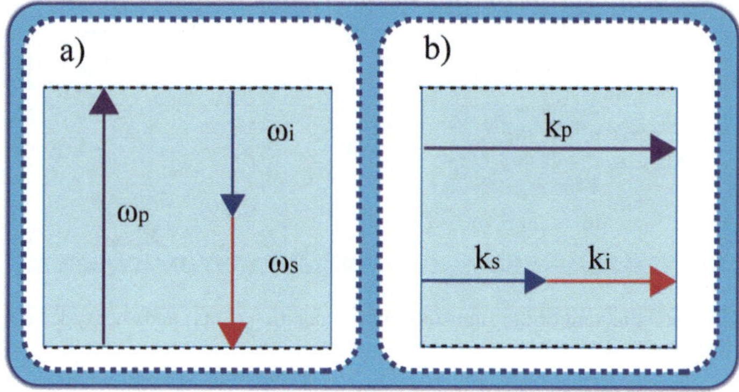

Fig. 1.6 Optical parametric process energy-level diagrams: (**a**) energy conservation; (**b**) phase matching

Another case is known as optical parametric generation (OPG). In this process, a pump beam is divided into two lower-energy beams, often called a signal beam and an idler beam. If the OPG is placed in a resonator, an optical parametric oscillator is built. An OPG is an OPA with a very high gain, so that a significant output power is generated even without any input signal [4]. Also, OPG has many characteristics, such as wavelength conversion and pulse width narrowing due to nonlinear propagation.

Furthermore, in nonlinear interactions, the process of phase pairing between interrelated waves is highlighted. On the other hand, in nanoscale nonlinear optical effects, there is no intervening between nonlinear optical interactions and wave phase coincidence derived from interacting optical fields. As a result, the wave vector dependence of optical fields is eliminated.

Also, regardless of the physical mechanism responsible for the coupling, phase matching is necessary for efficient coupling between waves or optical modes. Moreover, it is necessary for efficient nonlinear optical interactions. The necessary phase coincidence for the second-order nonlinear interaction is shown in Fig. 1.6.

Also, when a characteristic frequency changes sign, the corresponding wave vector at the phase match also changes sign. In a parametric process, the phase-matching condition is not automatically met, so this results in the conversion of optical frequencies. However, in a parametric process that excludes the conversion of energy from one optical frequency to another, it can be instantly satisfied. Furthermore, in any nonparametric process where there is an exchange of energy between the optical field and the medium, it is automatically fulfilled [5].

The second-order nonlinear optical response, also associated with first-order hyperpolarizability, physically involves doubling the frequency of a beam when it is incident on a material with nonlinear properties.

Furthermore, the search for materials with nonlinear optical properties is of great interest due to the large number of applications that can be designed through these properties. Therefore, the search for new materials for second-order nonlinear optical applications consists in the synthesis of highly efficient non-centrosymmetric molecules, and, at the same time, coupling these molecules into non-centrosymmetric bulk materials. It can result in the design of devices such as single crystal or polymer thin films for their use in specific applications.

In this search, it is necessary to measure and optimize the macroscopic and microscopic coefficients of the molecules of nonlinear materials. The properties of the microscopic coefficients are determined by the factor β associated with first-order optical hyperpolarizability, while the macroscopic properties are determined by the parameter $\chi^{(2)}$, associated with second-order nonlinear optical susceptibility [6].

Due to this, some techniques have been developed for the characterization of second-order nonlinear optical properties in molecules and bulk materials, as shown in Fig. 1.7.

First, it is necessary to define the characterization methods for the coefficients associated with the macroscopic and microscopic coefficients of the second-order nonlinear optical response. Microscopic or molecular characterization techniques are

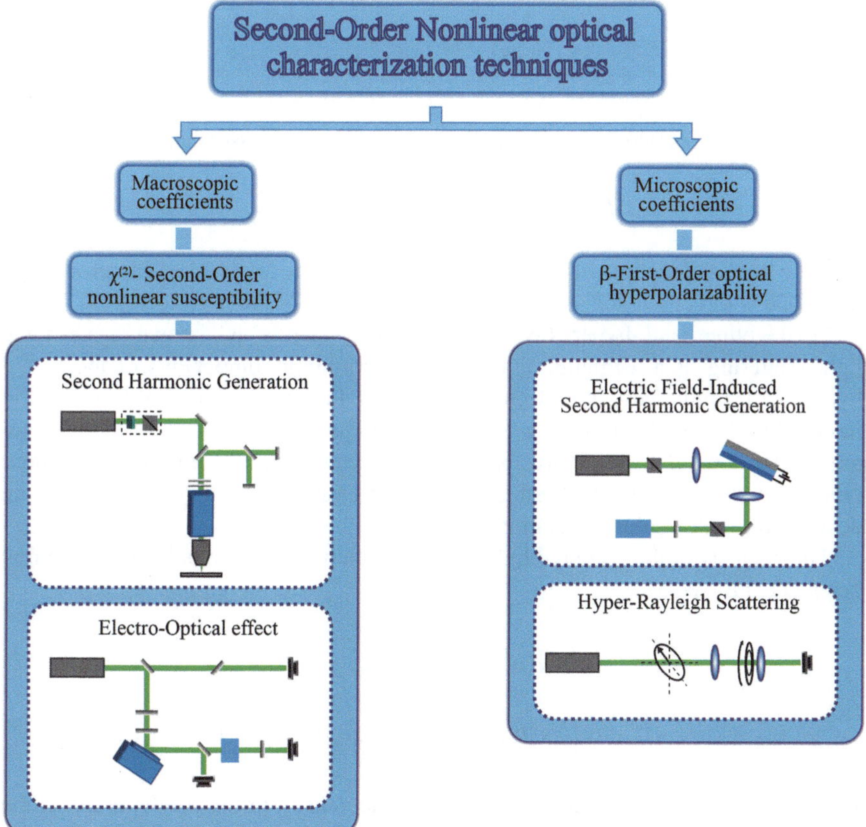

Fig. 1.7   Diagram of second-order nonlinear optical characterization techniques

used to determine the components of the first hyperpolarization tensor. The main techniques used are electric field-induced second harmonic generation, hyper-Rayleigh scattering, and solvatochromism. On the other hand, macroscopic characterization techniques associated with second-order nonlinear optical susceptibility are mainly second harmonic generation and linear electro-optical effect [7].

The technique of electric field-induced second harmonic generation is used mainly to obtain values associated with the first hyperpolarizability β and the dipole moment μ of the studied molecule. However, one of the main limitations of this technique is that it does not provide a direct measure of β. So, this technique is only applicable to dipolar molecules [8].

The electric field-induced second harmonic generation consists of inducing an external electric field to dipole molecules in solution. The mentioned molecules are oriented in the direction of the field, giving rise to a second harmonic generation. Moreover, the electric field-induced second harmonic generation allows determining the modulus and phase of the molecular hyperpolarizability $\Upsilon(-2\omega; \omega, \omega, 0)$. To obtain the first molecular hyperpolarizability $\beta(-2\omega; \omega, \omega)$ for a dipolar molecule, it is necessary to obtain the value of its second hyperpolarizability given by

$$\Upsilon(-2\omega; \omega, \omega, 0) = \frac{1}{5}\left[\Upsilon_{xxxx} + \Upsilon_{yyyy} + \Upsilon_{zzzz} + 2\left(\Upsilon_{xxyy} + \Upsilon_{xxzz} + \Upsilon_{yyzz}\right)\right] \quad (1.10)$$

The vector associated with the first hyperpolarizability tensor can be expressed as

$$\beta_z(-2\omega; \omega, \omega) = \beta_{zzz} + \beta_{zxx} + \beta_{zyy} \quad (1.11)$$

where z is the direction of the dipole moment of the molecule. Therefore, this technique is directly associated with the measurement of the β vector.

On the other hand, hyper-Rayleigh scattering, which is also known as harmonic light scattering, is a technique that allows determining components of the first β hyperpolarizability without molecular orientation. For this reason, unlike the electric field-induced second harmonic generation technique, measurements of the parameters of the first β hyperpolarizability can be carried out in molecules without permanent dipole moment. Furthermore, the measurements of this technique are developed on molecules in solution [9].

The intensity of the scattered light in the Υ direction for the z-polarized incoming beam is expressed as

$$I_Z^{2\omega} = g\left[N_S\langle\beta_{zzz}^2\rangle_s + N_P\langle\beta_{zzz}^2\rangle_p\right]e^{-\varepsilon(2\omega)lN_P}I_\omega^2 \quad (1.12)$$

$$I_x^{2\omega} = g\left[N_S\langle\beta_{zxx}^2\rangle_s + N_P\langle\beta_{zxx}^2\rangle_p\right]e^{-\varepsilon(2\omega)lN_P}I_\omega^2 \quad (1.13)$$

where N_s is the density of the solvent, N_p is the density of the solute, $\varepsilon(2\omega)$ is the molar extinction coefficient of polymer at 2ω frequency, and l is an effective optical path.

The value of the parameter g involved is obtained through calibration with a solvent of known β value. The g factor considers local fields and geometric factors, and the averages are given by

$$\langle \beta_{zzz}^2 \rangle = \frac{1}{7}\sum_i \beta_{iii}^2 + \frac{6}{35}\sum_{i\neq j}\beta_{iii}\beta_{ijj} + \frac{9}{35}\sum_{i\neq j}\beta_{ijj}^2 + \frac{6}{35}\sum_{i,j,k,\,\text{cyclic}}\beta_{iij}\beta_{jkk}$$

$$+ \frac{12}{35}\beta_{ijk}^2 \tag{1.14}$$

$$\langle \beta_{zxx}^2 \rangle = \frac{1}{35}\sum_i \beta_{iii}^2 + \frac{2}{105}\sum_{i\neq j}\beta_{iii}\beta_{ijj} + \frac{11}{105}\sum_{i\neq j}\beta_{ijj}^2 - \frac{2}{105}\sum_{i,j,k,\,\text{cyclic}}\beta_{iij}\beta_{jkk}$$

$$+ \frac{8}{35}\beta_{ijk}^2 \tag{1.15}$$

In this technique, the components of the β tensor from the voltage of the scattered light at a harmonic frequency versus the square of the intensity of the beam, measured for different polarization configurations, are obtained. On the other hand, one of the main drawbacks of the hyper-Rayleigh scattering technique is that some other nonlinear optical processes can contribute to the measured signal. Therefore, it is convenient to use this technique under the femtosecond regime since it allows separating the photons of the harmonic generation from the others.

As can be seen, molecular hyperpolarizability is determined through microscopic characterization techniques or at the molecular level. The main characterization techniques for the measurement of molecular hyperpolarizability are the generation of the electric field-induced second harmonic generation and the hyper-Rayleigh scattering. The electric field-induced second harmonic generation oversees measuring the product of the dipole moment (μ) and β, while the hyper-Rayleigh dispersion measures β directly at a specific fundamental wavelength. So, in the electric field-induced second harmonic generation technique field to measure the optical parameters $\mu\beta$, a large electric field applied to the nonlinear optical system induces a general dipole alignment, such that it allows the second harmonic generation in an intense way, unlike for the hyper-Rayleigh technique in which the application of an electric field to the sample is not necessary. The hyper-Rayleigh technique produces the orientation of the molecular tensors through the signal derived from the second harmonic generation. The hyper-Rayleigh technique has the advantage of allowing the measurement of molecular hyperpolarizability in dipolar and nondipolar molecules. However, due to the oscillations associated with the density and molecular orientation of the material in solution, the intensity generated through the second harmonic generation is weak, and on the other hand, since the value of the molecular optical hyperpolarizability also depends on the incident wavelength. Therefore, to determine the relevant β value, it is necessary to measure over a range of wavelengths to establish an operating wavelength for a specific device.

Another method not mentioned above is solvatochromism. This technique is rarely used because the molecular hyperpolarizability measurement values are approximate. In addition, the measurement values of the parameters depend on other physical factors of the sample, such as the solvent in the suspension and the radius of the cavity [10].

The second harmonic generation is a nonlinear second-order optical effect that is characteristic of being a type of frequency doubling. This process consists of the interaction of a nonlinear material with two photons with the same frequency. This interaction results in the combination of the photons and the generation of a new photon with twice the energy of the initial photons. Estop is also equal to twice the frequency and half the wavelength, which preserves the coherence of the drive. The second harmonic generation technique is considered as a special case of frequency sum generation and specifically of harmonic generation [11].

On the other hand, second harmonic generation is not possible in non-centro-symmetric media like other nonlinear second-order optical phenomena. That is why, unlike third-order nonlinear optical effects, in second-order nonlinear optical effects the study of the surfaces and interfaces of the optical systems studied is of utmost importance.

Optical second harmonic generation is a technique that provides fast, second-order electronically sourced susceptibility $\chi^{(2)}(-2\omega; \omega, \omega)$ at a measurement frequency ω. The second harmonic generation technique focuses on measuring twice the incident frequency 2ω given an electric field $E(\omega)$. The above is done in order to establish a relationship with the surface of the optical system such that the induced dipole of the second harmonic per unit volume $P_{\mathrm{NL}}^{(2)}(2\omega)$ can be expressed as

$$E(2\omega) \propto P_{\mathrm{NL}}^{(2)} = \chi(2)E(\omega)E(\omega) \tag{1.16}$$

where $\chi^{(2)}$ is known as the second-order nonlinear optical susceptibility tensor. Furthermore, the generated electric field $E(\omega)$ and the corresponding second-order nonlinear optical susceptibility reveal information about the orientation of molecules on a surface. This technique is also highlighted for exhibiting information about the interfacial analytical chemistry of the surfaces and the chemical reactions at the interfaces.

So, the interaction of a light beam with a nonlinear optical medium can be described in terms of nonlinear polarization. Hence, it is possible to define the effective susceptibility through

$$P_{\mathrm{NL}} = \chi_{\mathrm{ef}}{}^{E} \tag{1.17}$$

where

$$\chi_{\mathrm{ef}} = \chi^{(1)} + \chi^{(2)}E \tag{1.18}$$

The relation between the intensities of the second harmonic signal and the incident pulse signal is a quadratic relationship as shown below:

$$I_{2\omega} \propto I_{\omega}^{2} \tag{1.19}$$

Then, the intensity as a function of the length of the medium is given by

$$I_{2\omega}(L) \propto \left(\frac{\omega L \chi_{\text{ef}}^{2}}{n_{2}\omega c}\right) I_{\omega}^{2} \sin^{2}\frac{L\Delta k}{2} \tag{1.20}$$

where $\omega\ m$ is the angular frequency of the incident wave, χ_{ef} is the effective second-order susceptibility of the material, $n_{2\omega}$ is the refractive index of the medium at frequency 2ω, and Δk is the difference of the magnitudes of the wave vectors of the incident beam and the second harmonic generation beam. The difference between the magnitudes of the wave vectors is

$$\Delta k = k_{2\omega} - 2k_{\omega} \tag{1.21}$$

where the wave vector of the incident beam is represented by

$$k\omega = \frac{\omega n_{\omega}}{c} \tag{1.21}$$

Also, the wave vector of the second harmonic generation is

$$k_{2\omega} = \frac{\omega n_{2\omega}}{c} \tag{1.22}$$

Finally, the result of the phase shift between the second harmonic waves that are generated through the nonlinear medium of length, L, and that propagate collinear with the incident wave is

$$\sin^{2}\frac{L\Delta k}{2} \tag{1.23}$$

On the other hand, the electro-optical effect originates due to the change of the optical properties of a material in response to an electric field that varies compared to the frequency of light. The application of an electric field on a nonlinear optical system results in the dislocation of charges, which produces dipoles or reorients existing ones. The electric field induces or modifies an anisotropy in the nonlinear optical system such that the application of the static field modifies the dielectric tensor and the refractive indices of the material is possible. Likewise, the ellipsoid of the refractive indices is the cause of defining the type of anisotropy in the medium. Furthermore, if the applied field is oscillating, in addition to the modification of the optical anisotropy, mechanical resonances synchronized with the mechanical modes of the solid can occur. On the other hand, the linear electro-optical effect as a function of the ability of a nonlinear medium to modify its refractive index through an incident external electric field is manifested. Such modification can be defined proportionally to the intensity of the external field. Through this principle, there have been developed some techniques whose main objective is to measure the electro-optical susceptibility of thin films and crystals [12].

As described above, some methods to characterize materials associated with nonlinear optical susceptibility are the electro-optical effect and the second harmonic generation. Through these techniques, the coefficients of the generation of the second harmonic and the electro-optical effect are affected. By the enhancement of the resonance, a nonlinear optical system can present optical absorption in the wavelength of the main harmonic or the second harmonic of the incident light. Concerning second harmonic generation, the nonlinear optical susceptibility can be determined by measuring the polarized light of the second harmonic. The light coming from the sample can be analyzed from the different angles of incidence of the beam and by comparing the efficiency of the second harmonic of a quartz wedge. In such a way, the second-order nonlinear optical susceptibility can be determined through the analysis of the second harmonic pattern obtained. On the other hand, techniques based on ellipsometry and two-slit interference modulation to measure the electro-optic coefficients induced by the Pockels effect have been described. Similarly, the values of the electro-optic coefficient, such as molecular hyperpolarizability, change as a function of wavelength by virtue of the frequency dependence of the dielectric permittivity of the sample.

1.3 Experimental Techniques in Third-Order Optical Nonlinearities Evaluations

Through the following roadmap, a review of the process of study and characterization of materials is described to design applications. The study process focuses on the optical properties related to third-order nonlinear optical susceptibility. Such third-order nonlinear optical properties of the material include nonlinear saturable absorption, reverse saturable absorption, two-photon absorption, multiple photon absorption, nonlinear refraction, and nonlinear scattering. Therefore, it is important to specify the characterization and experimental methods used to measure optical nonlinearities.

For the synthesis and production of materials, two methods are shown in Fig. 1.8. First, is the bottom-up method, which consists of accumulating atom by atom to form nanostructures. The control in bottom-up processes is often much greater than in top-down methods, which involves the production of high-quality nanoparticles. Three of the most common methods are atomic shell deposition, physical vapor deposition, and chemical vapor deposition.

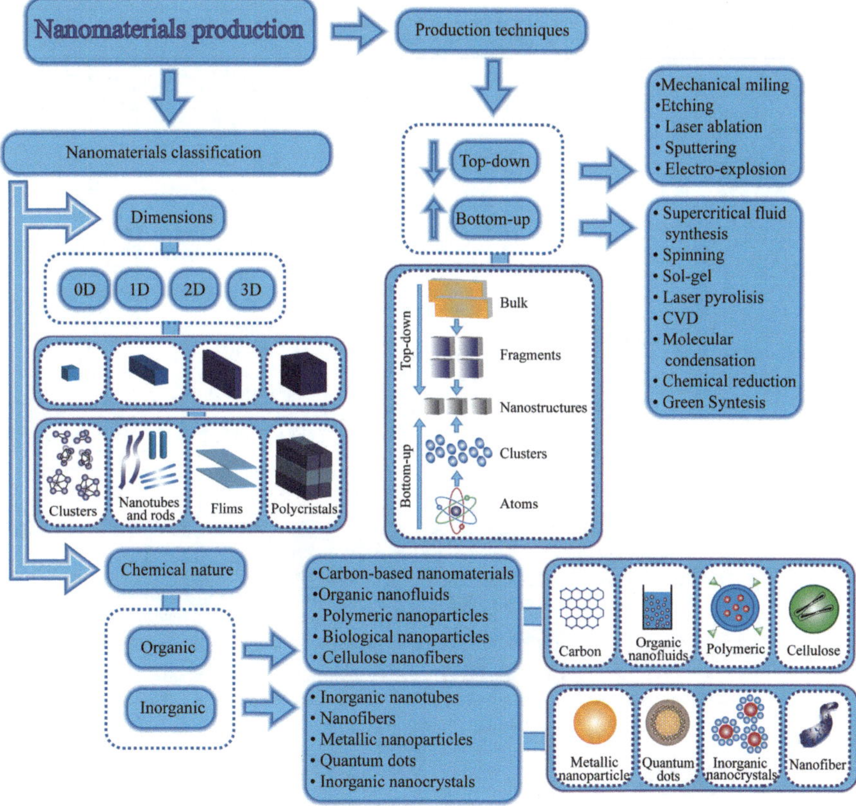

Fig. 1.8 A roadmap for the nanomaterials production

On the other hand, the top-down method consists of reducing bulk materials to nanometric sizes. To achieve material reduction, it is necessary to take a large bulk material and break it down using different types of energy, such as thermal, mechanical, or chemical, to modify the material into structures in the nanoscale range. One of the advantages of top-down production methods is the fact that they can be used for mass manufacturing. However, the disadvantage is that they lack control due to the use of various chemical compounds. Lithography and engraving methods fall under the category of top-down methods. These methods are used to remove material and model nanoparticles to obtain a specific surface. Also, grinding, shearing, and ball milling are traditional physical methods for grinding bulk materials into nanoparticles.

The nonlinear absorption coefficient and nonlinear refractive index are optical parameters that depend on the components of the third-order nonlinear optical susceptibility tensor $\chi^{(3)}$. To gain an understanding of the optical parameters, there

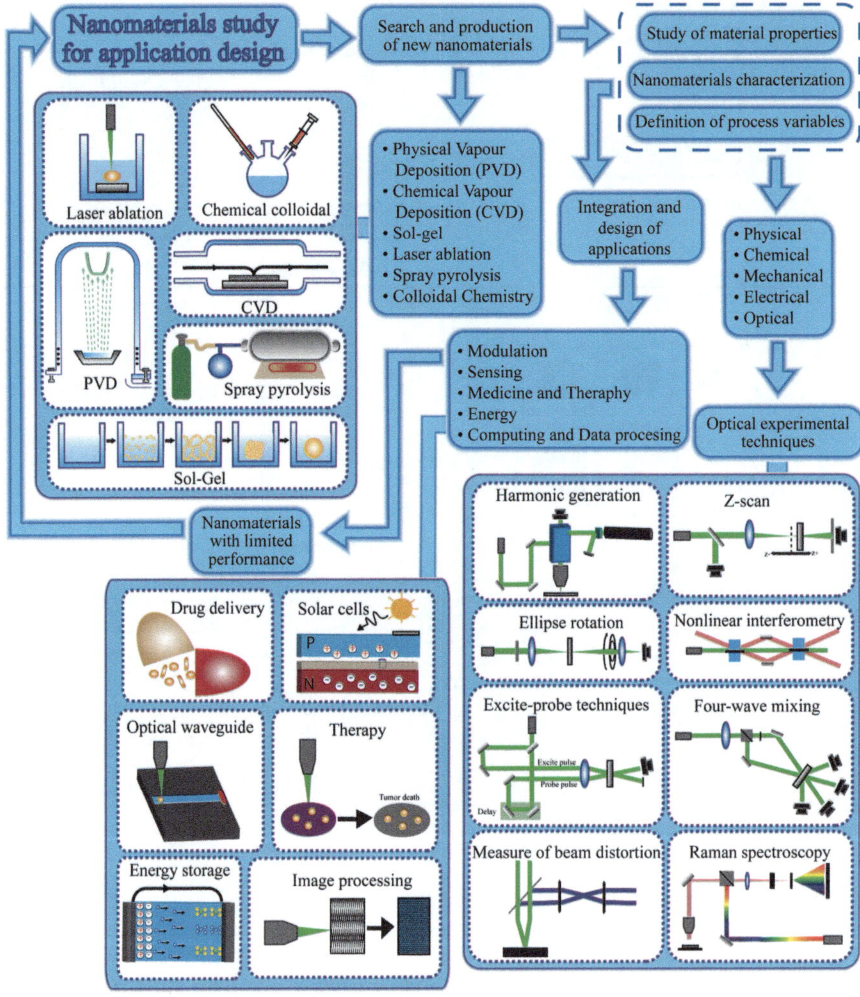

Fig. 1.9 A roadmap for the study of 2D materials for photonics applications

are some experimental methods that evaluate this contribution as shown in Fig. 1.9. The techniques developed for the measurement of the parameters are the third harmonic generation, z-scan, coupling of two optical waves, ellipse rotation, four-wave mixing, Raman spectroscopy, measurement of beam distortion, etc. So, next, an analysis of the various measurement techniques of the optical parameters and the contribution of the third-order optical susceptibility $\chi^{(3)}$ was carried out. In addition, some of these techniques can identify the physical mechanisms of optical nonlinearity exhibited by the samples tested.

The third harmonic generation consists of generating a laser beam with a third of the wavelength from an excitation beam. That is, the third-order coefficient gives rise in this case to the generation in the middle of a wave at three times the frequency of the incident wave such that three photons become a single photon with three times the original frequency. Various aspects of the nonlinear optical behavior of nanoparticles related to the generation of third harmonics from a variety of materials have been reported in the literature. So, from these studies, it is possible to determine applications of this nonlinear optical effect. The third harmonic generation technique is one of the most used techniques in the study of the contribution of third-order susceptibility due to its simplicity. In the same way, it is a very suitable technique in the case of the study of fluids because it helps to understand the nonlinear electronic polarizability of large molecules associated with dilute solutions. There are some other useful techniques in the characterization of fluids; however, it is necessary to perform additional experiments to determine and correct the influence of vibrational, rotational, and intermolecular reorganization movements. This is regarding the parameters related to the movement between the fluid molecules are subject to electronic scattering effects. These factors result in the poor functionality of other methods in the optical measurement of fluids for the evaluation of hyperpolarizability models.

Otherwise, one of the restrictions of the third harmonic generation method in solutions is the fact that liquids do not have strong absorption at fundamental and harmonic frequencies. In addition, the intensities due to the third harmonic generation method appear in a complex way as interferences between the nonlinearity discontinuities of the materials. However, it has the advantage that the phase mismatch attributed to the nonlinear polarization wave and the harmonic free wave is a linear function of the transverse displacement. Thus, the behavior allows the simple determination of the phase mismatch.

The third harmonic generation technique produces nondegenerate signals in optical frequency and space. Therefore, it is possible to design applications compatible with low-cost electronic components, which offers advantages over other typical means. The third harmonic generation enables a variety of optical image processing applications because it is an inherently ultrafast electronic process. Applications based on the generation of the third harmonic produce a limited response time due to the duration of the ultrafast pulses. On the other hand, the phase unpaired nature of third harmonic generation offers the possibility of designing optical processing applications with large spectral and angular bandwidths. Also, applications derived from the third harmonic generation process offer the critical advantage over the use of other nonlinear optical processes, such as the Kerr effect, because they are not susceptible to dispersion at the fundamental wavelength. The advantage is due to the signals originating from the third harmonic generation process that have facilities to be spatially and spectrally filtered. Thus, through optical 2D image processing, the applications assigned to the third harmonic generation method are expanded. It has been shown that, through the generation of the third harmonic, a configuration based on the compact Fourier transform can be used. The configuration is characterized through a diffractive optical element in order to generate input beams that establish a spatial and temporal superposition [13].

Also, one of the main advantages of the Raman spectroscopy technique is the ability to eliminate the incoherent fluorescence that causes the darkening of the desired spectrum. Nonlinear effects can exhibit background signals in the Raman technique. In addition, they can alter or obscure the resonances due to the active Raman modes of the sample. Thus, nonlinear background signals are the result of frequency-independent contributions to the optical coefficients. For this reason, the incident laser intensities produce a false structure, which can be mistaken for weak Raman modes.

Moreover, the z-scan technique appears to be a very simple and effective technique in measuring the real and imaginary parts of Kerr's optical susceptibility. In z-scan measurements, the sample moves along the direction of propagation of a highly focused beam, and the variation of the far-field intensity is used to determine the nonlinear optical response. However, the origin of the optical susceptibility also seems to be one of the problems in the implementation of the technique. So, problem-solving is used in the measurement of global Kerr optical susceptibility. Then, this allows emerging thermal mechanisms that result in physical changes related to the expansion of the measured materials. Furthermore, it is known that, in the case of solutions, there is an important contribution to Kerr susceptibility from the rotation of molecules unlike the third harmonic generation technique that allows the fast electronic part of third-order optical susceptibility to be measured because the measured harmonic field mimics the fundamental field, oscillating with the same frequency, such that it is possible to determine the highest value of the response time. The differences between the parameters measured through the measurement techniques are associated with the important rotational contributions of the z-scan technique. However, despite the simplicity of the experimental setup and in the data analysis of the z-scan technique, a high-quality Gaussian beam is necessary for absolute measurements. Thus, the fluctuations in the detected signal from the inclinations of the sample during the measurement are derived. This can cause the beam to fall outside the far-field aperture. Therefore, it is not possible to use such a technique to measure off-diagonal elements of the susceptibility tensor, except when a second nondegenerate frequency beam is used.

Besides, nonlinearities excited in a short time are of special interest since with this particularity it is possible to use them in ultrafast optical communication, information processing, and storage devices and systems. So, to obtain these characteristics it is important to use techniques that involve generating short pulses.

The degenerate four-wave and z-scan mixing methods have been widely used in the nano- and picosecond pulse width ranges. These methods, as mentioned above, are widely used to determine third-order nonlinear optical susceptibility. However, the disadvantage is that it is only possible to indirectly demonstrate the presence of a fast electronic component of optical nonlinearity. Electronic components in nonlinearities have a characteristic settling time in a time range of femtoseconds. Thus, this causes the isolation of the electronic component from other nonlinearity mechanisms that have short settling times compared to the widths of the test pulses.

Another technique used is the nonlinear ellipse rotation technique, which consists of the irradiance of a single beam of elliptically polarized wave. The elliptically polarized beam with left and right circularly polarized components results from the incidence of a linearly polarized wave on a quarter-wave plate. The nonlinear ellipse rotation technique can determine only one component of the tensor, but it has the limitation that it cannot measure the total third-order nonlinear susceptibility. Thus, a convenient solution is to combine the advantages of z-scan and nonlinear ellipse rotation to measure the tensor component and third-order nonlinear susceptibility in a simple and sensitive way.

The multiwave mixing consists of the interaction of electromagnetic fields to produce a new one. Understanding the coupling of optical waves gives rise to considering the individual interactions of electromagnetic fields through a dielectric medium. The incoming electromagnetic field induces an oscillating polarization in the dielectric medium. The input electromagnetic field is re-irradiated with a phase shift determined by the damping of the individual dipoles, which is also due to Rayleigh scattering. The interaction between the fields will also induce the polarization of the dielectric medium and the interference of the two waves will cause harmonics in the polarization at the frequencies. Thus, frequencies can act as new source fields, which can produce numerous interactions.

Thus, one of the promising methods for time spectral resolution is the two-wave mixing technique to determine the nonlinear third-order optical susceptibility of materials. The two-wave mixing technique allows time resolution to determine the time to establish nonlinearity in the femtosecond range of pulse widths and separate the contributions of various nonlinearity mechanisms. It also has a high sensitivity compared to other techniques. Furthermore, the technique is sensitive to the variation of the pulse parameters with the nonlinearities of the medium. It also highlights the fact that it is a simple technique to implement. On the other hand, the technique has the disadvantage that it requires a high stability of the spectrum of the pulses that are generated. This is in addition to knowledge of the geometry of the axes of the beams in the sample. In another way, there are errors in the technique due to the intensity distribution in the cross section and the asymmetry of the phase increase.

One of the main applications of multiwave mixing is based on phase conjugation. Then, through the phase conjugation derived from the coupling of optical waves, the design of intracavitary elements of laser systems is possible. Through an intracavitary element, it is possible to eliminate the distortions introduced during the laser process as they pass through the cavity again. The application can be optimized using crystals. The input wave is scattered through the crystal and reflected internally by the crystal faces cut at the phase-matching angle. These reflected waves then act as pump waves in optical wave coupling. Considering the third-order nonlinear effects, optical wave mixing can be used for holographic imaging since the interaction of multioptical waves acts as a refractive index grating in optical media. Thus, due to the nonlinear interaction between the dielectric medium and fields, the network modulates the intensity separately from the third-order susceptibility. The information is stored in the interference pattern through the input beam reflected from an object. Subsequently, the original wave is reconstructed by diffracting the third wave of the interference pattern. Also, the Fourier transform of an image can be obtained through an image reflected through the focal plane of a lens. On the other hand, the Fourier product is obtained through a fourth wave with induced polarization coming from the focus of three waves that contain spatial information, derived in turn from the reflection of a nonlinear medium in the focal plane of a lens such that it is possible to determine the convolution of two waves in combination with a third, through Fourier analysis. Then, spatial correlations or instantaneous convolutions can be performed [14].

The degenerate four-wave mixing is based on various interactions between light and matter in which three incoming waves in the material generate a fourth frequency wave. The strength of the incident fields and the polarization efficiency of the material define the magnitude of the polarization. The polarization by the third-order nonlinear susceptibility $\chi^{(3)}$ is indicated. Also, the magnitude of third-order susceptibility is highly dependent on the incident and induced frequencies of the four-wave degenerate mixing technique. That is why, if the simple or combined frequencies are in resonance with the frequencies, the third-order optical susceptibility of the material will exhibit a high value.

Thus, the four-wave degenerate mixing technique can measure different components of the $\chi^{(3)}$ tensor with the advantage of using an experimental model for isotropic materials. Also, the beams used need not be perfect Gaussian. However, one of the disadvantages of using this technique is the fact that generally only the modulus of $\chi^{(3)}$ can be measured. So, a more complicated experimental model is necessary to extract the real and imaginary parts of $\chi^{(3)}$. That is why it is necessary to perform various measurements to unravel the underlying physical mechanism by varying parameters such as irradiance and polarization state, or different measurement methods.

On the other hand, the third-order nonlinear optical effect called self-focusing is defined as a decrease and increase in the convergence of high-power radiation through a nonlinear medium. Through such an effect, a focusing spatial distribution is possible, resulting in the robustness of the waves. So, the main purpose of the self-focusing effect is the directed transmission of the concentrated radiation and the focusing of the radiation at a single point. So, the self-focusing effect is optimal for the design of applications that involve radiation energy, in addition to thermal effects such as the heating of matter, control of destruction processes, and particle acceleration.

Therefore, optimization of systems used for super-resolution applications is possible through the configuration of the incident bomb beam as a surface plasmon polariton. The said polariton results in the modification of its enhancement electromagnetic field, in which the nonlinear interaction force increases. Through this technique, the nonlinear optical resource is used to transform the information stored into propagation waves. Thus, the possibility of partial phase interactions of electromagnetic fields through a metal–dielectric interface is also explored to assign spatial information stored in evanescent waves [15]. Besides, models have been implemented based on the simulation of the propagation of high-energy laser pulses through nonlinear optical materials using focused beams. Ultrafast nonlinearities of the optical Kerr effect exhibited by the materials are considered in the simulations. Equally, the consideration of the absorption and refraction coefficients is important. Thus, the exhibited nonlinearities accumulate in the laser pulses over time. On the other hand, it is also important to consider the thermal mechanisms induced by laser irradiation in the samples, such that the computational simulations pose a small error bar [16]. The induced Kerr optical effect, as a function of third-order nonlinear optical susceptibility, results in the dependence of the material's complex optical index on the wave intensity. So, by modifying the optical absorption in a medium through light, materials exhibiting large nonlinear optical coefficients are great candidates for the design of fully optical signal processing devices, such as ultrafast switches. Thus, the application designs involve materials with a high nonlinear response in a specific spectral range, in addition to a low linear absorption. It is also important to consider the fast response times of the materials.

On the other hand, the ultrafast optical Kerr effect has been established as a tool for recording the ultrafast dynamics of liquids with high time resolution characteristics. For this, it is also necessary to consider physical–chemical characteristics such as the vibrational and diffusive orientation movement since these characteristics are responsible for a significant fraction of the dynamics. Therefore, the optical exploration through the optical Kerr effect of the dynamic behavior of fluids allows the control of discrete variables associated with liquid physicochemical changes. Furthermore, the dynamics of the proposed nanofluids favor the exploration of signals through a liquid. Due to the control of the variables by the optical nonlinearities of the samples, it is possible to obtain a dynamic exploration, a difference from hybrid liquid mixtures that, due to their size and dynamics, Kerr nonlinearities can lead to the design of quantum systems responsible for causing phase changes dependent on optical irradiance. Quantum optics has even been shown to generate photon nonlinear optical methods for the development of optical switches. Furthermore, the influence of optical waves on the mechanical dynamics of nanofluids can explore different atmospheres [17].

1.4   Nonlinear Optical Effects Represented by the Transmitted Beams in the Two-Wave Mixing

To describe the behavior of nonlinear optics, it is important to understand the transmission of light waves through nonlinear optical media. The response of the propagation of electric fields through nonlinear media can be described by the following wave equation:

$$\nabla^2 E_{\pm} = -\frac{n_{\pm}^2 \omega^2}{c^2} E_{\pm} \tag{1.24}$$

where E_+ and E_- represent the circular components of the right and left electric fields, respectively; the optical frequency corresponds to ω, the refractive index dependent on irradiance is n, and c is the speed of the light.

The scattering behavior of the thin thickness nonlinear grating can be defined through Raman–Nath diffraction because this phenomenon produces a large number of diffracted waves. In Raman–Nath scattering, the light beam interacts with the diffraction grating, and due to the interaction with the nonlinear medium, the refractive index of the medium varies with the period. The incident light beam is diffracted from this grating, which modifies the refractive index.

Through the self-diffraction derived from the interaction of two waves in an optical medium with a Kerr nonlinearity, the amplitudes and polarizations of the transmitted and self-diffracted waves can be calculated.

The birefringence simulation allows calculating the vector amplitude represented by the product of the incident beam and the Jones matrix that represents the transmittance of the sample. It is set to define the phase increments of the bias components. In the same way, the matrix is diagonal due to the representation of the circular polarization components $E_{1+}, E_{1-}, E_{2+}, E_{2-}$. The amplitudes of the diffraction orders of the transmitted waves can be expressed as [18]

$$E_{1\pm}^{(0)} = \left[J_0\left(\psi_\pm^{(1)}\right) E_{1\pm} + iJ_1\left(\psi_\pm^{(1)}\right) E_{2\pm}^{(0)} \right] \times \left(-i\psi_\pm^{(0)}\right) \tag{1.25}$$

$$E_{2\pm}^{(0)} = \left[J_0\left(\psi_\pm^{(1)}\right) E_{2\pm} + iJ_1\left(\psi_\pm^{(1)}\right) E_{1\pm}^{(0)} \right] \times \left(-i\psi_\pm^{(0)}\right) \tag{1.26}$$

where a property of the Bessel functions is defined; similarly, the phase increments of the polarization components can be written as

$$\psi_\pm(x) = \psi_\pm^{(0)} + \psi_\pm^{(1)} \cos \frac{2\pi x}{\Lambda} \tag{1.27}$$

$$\psi_\pm^{(0)} = \frac{4\pi^2 d}{n_0 \lambda} \left[A\left(|E_{1\pm}|^2 + |E_{2\pm}|^2\right) + (A + B)\left(|E_{1\mp}|^2 + |E_{2\mp}|^2\right) \right] \tag{1.28}$$

$$\psi_\pm^{(1)} = \frac{4\pi^2 d}{n_0 \lambda} \left[A E_{1\pm} E_{2\pm}^* + (A + B) E_{1\mp} E_{2\mp}^* \right] \tag{1.29}$$

where $\Lambda = \lambda/2 \sin \theta$ and represents the period of the grating induced in the nonlinear medium.

On the other hand, the basic principle of wave coupling is based on the interaction of coherent beams that interfere in the medium to induce changes in the optical properties dependent on the irradiance of the medium. Also, the phenomenon depends on the diffraction grating, in which the incident waves are diffracted at the outlet of the medium. Optical wave coupling has the utility of analyzing the intensity and polarization dependencies of the interacting waves in an optical material. It also has the purpose of evaluating the possibilities of controlling the polarization and optical amplification of pulses.

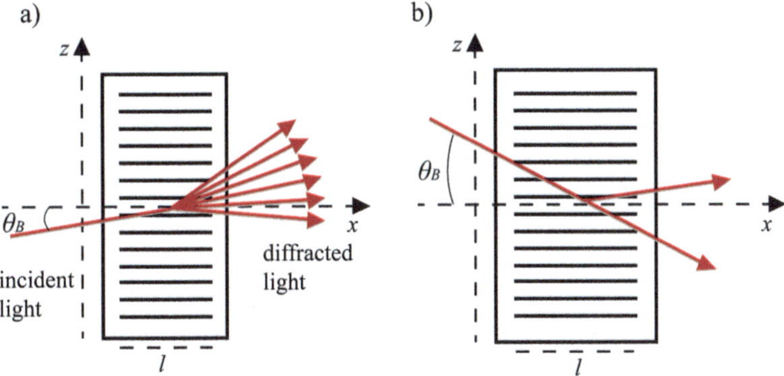

Fig. 1.10 (**a**) Diagram of Raman–Nath diffraction; (**b**) Bragg diffraction diagram

Wave coupling is based on the interaction of coherent waves in a nonlinear medium through third-order susceptibility. The optical wave mixing process can be generated in any optical medium because optical media have third-order non-linearities difference of zero. If at least two high-intensity waves interact, it is possible to generate nonlinear effects through the medium. The optical wave mixing process can be carried out efficiently if there is phase overlapping, which is achieved by using appropriate angles between the phase propagation.

The effects related to the mixing of optical waves involve the generation of laser-induced diffraction gratings on the material. The induced gratings can be of trans-mittance effects related to the modulation of the nonlinear absorption of two photons. Also, the grating effects per phase are due to the modulation of the refractive index due to the optical Kerr effect. The diffraction gratings are a function of the physical mechanism responsible for the modification of absorption and refraction.

The coupled wave theory is based on vector equations that define the behavior of the steady-state amplitudes of coupled waves. In addition, it allows observing the dependence of the coupling of waves and the phase adaptation of the polarization of the waves. The system of vector equations is based on the nonlinear coupling effects for wave mixing by describing the changes in the polarization of the beams.

In the specific case of phase mismatches, the Bragg and Raman–Nath diffraction theories observed in Fig. 1.10 should also be addressed. In this way, it is possible to analyze the coupling of waves with different polarization.

The following system of equations proposes wave coupling solutions with linear and circular polarization for the analysis of vector wave interaction [19]:

$$E_1(z) = e^{-i\frac{\Delta k_{0x}}{2}}E_1(0) \tag{1.30}$$

$$E_{2m}(z) = e^{-i\beta z}\{E_{2m}(0)\cos h\gamma_m z$$

$$+\frac{i}{\gamma_m}\left|[(\Delta k_0 - \Delta k_{1m})E_{2m}(0) - \Delta k_{1m}E_{3m}(0)]\sin h\gamma_m z\right\} \tag{1.31}$$

$$E_{3m}(z) = e^{-i\beta z}\{E_{3m}(0)\cos h\gamma_m z + \frac{i}{\gamma_m}$$

$$\times \left|[(\Delta k_0 - \Delta k_{1m})E_{3m}(0) - \Delta k_{1m}E_{2m}(0)]\sin h\gamma_m z\right\} \tag{1.32}$$

Besides, it is possible to consider the theory of three waves coupled to the study of the nonlinear behavior of metallic anisotropic materials. Due to the metal particles in the material, it is necessary to consider the refraction of the resulting light.

Some of the applications based on the optical Kerr effect are focused on photonic devices controlled through Bragg gratings. The main functional character of the said devices is based on third-order optical nonlinear of an isotropic medium. Therefore, the selection of an available material, with optical and tunable nonlinearities, is key in the development of photonic control devices. Thus, the interaction of the optical Kerr effect is fundamental in the tuning and storage of encoded information through a state of the optical field. Equally, it is possible to establish the development of optical switches through a tunable Bragg grating [20].

In order to determine the components of the third-order susceptibility tensor derived from the associated nonlinearities of an anisotropic metallic material, the incidence with respect to the electric field must be taken into account. The description of the equation needs to consider different wave mixing interactions for the anisotropic metallic material. The interaction at normal incidence implies the three components of the susceptibility tensor, as well as the interactions for the determination of, so, the mentioned equation can be described as [21]

$$P_{\mathrm{NL}}^{(3)}(\theta,\omega) = |E(\omega)|^2 E(\omega)\chi_{\mathrm{eff}}^{(3)} \tag{1.33}$$

where

$$\chi_{\mathrm{eff}}^{(3)} = 3\left[\left(\chi_{1111}^{(3)}\sin^3\theta + \frac{3}{2}\chi_{1133}^{(3)}\sin 2\theta\cos\theta\right)\hat{j}\right.$$

$$\left. + \left(\frac{3}{2}\chi_{1133}^{(3)}\sin 2\theta\sin\theta + \chi_{3333}^{(3)}\cos^3\theta\right)\hat{k}\right] \tag{1.34}$$

is the effective third-order susceptibility of the material to an inclination θ between the material and the incident electric field, after considering the refraction of light.

The refractive index depends on the wavelength of the light through the medium. However, when speaking of nonlinear optics, the refractive index also depends on the intensity of the irradiated light on the medium. To describe the dependence of the material on the intensity of the field, it is necessary to consider the polarization of the medium. Thus, the polarization of the medium can be represented as the power dependence of the electric field. The change in refractive index described is also known as the optical Kerr effect. Then, it is possible to describe the vector Kerr effect from the approximation of the dependence of the nonlinear refractive index of the irradiance:

$$n_{\pm}^2 = n_0^2 + 4\pi\left(A|E_{\pm}|^2 + (A+B)|E_{\mp}|^2\right)$$ (1.35)

where n_0 is the weak field refractive index. $A = X^{(3)}{}_{1122}$ and $B = X^{(3)}{}_{1212}$ are the independent components of the third-order susceptibility tensor $\chi^{(3)}$.

The third-order nonlinear susceptibility $\chi^{(3)}$ can be calculated from the nonlinear refractive index n_2, and the nonlinear absorption coefficient β, considering

$$\chi_{1111}^{(3)} = \frac{n_0 c}{7.91 \times 10^2} n_2 + i \frac{n_0^2 c\lambda}{\pi^2}\beta$$ (1.36)

Therefore, the intensity-dependent index of refraction can be alternatively simplified as

$$n = n_0 + n_2 I$$ (1.37)

where I refers to the intensity of the optical field. Moreover, the linear and nonlinear refractive indices are related to the linear and nonlinear susceptibilities:

$$n_0 = \left(1 + \chi^{(1)}\right)^{1/2}$$ (1.38)

$$n_2 = \frac{3\chi^{(3)}}{4n_0}$$ (1.39)

When an electric field is applied parallel to the incident beam, the light propagates as two polarized waves orthogonally parallel to the other axes, such that different refractive indices are experienced in the system, as shown in Fig. 1.11.

The phenomenon of optical interference refers to the superposition of two or more electromagnetic waves. For the interference phenomenon to occur, the optical beams must be coherent and have the same wavelength and polarization. The interference of optical beams on the same plane results in a pattern of light intensity due to the distribution where the intensity is different from the individual sum of intensities of the beams due to its wave nature.

Fig. 1.11 Representation of the change in refractive index by optical Kerr effect

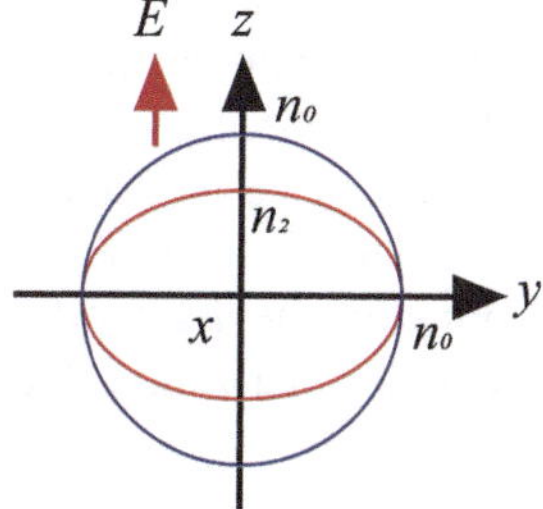

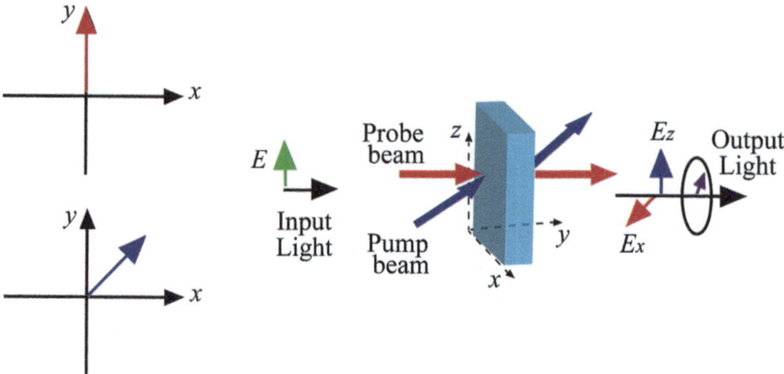

Fig. 1.12 Optical Kerr effect scheme by optical two-wave mixing

The optical Kerr effect consists of the rotation of the plane of polarization of light in a medium when it is subjected to an electric field. So, by applying an electric field to an isotropic material, the refractive index will change due to the optical Kerr effect such that a light beam with polarization parallel to the electric field experiences a smaller refractive index, as shown in Fig. 1.12.

The resulting intensity is obtained from the sum of the assumed fields, and in the case of two beams the intensity is given by

$$I = |E_1 + E_2|^2 \tag{1.40}$$

Considering then the superposition and interference of harmonic waves E_1, E_2, the superposition of the waves produces an intensity distribution as follows:

$$I = 2I_0[1 + \cos(2kx\sin(\theta))] \tag{1.41}$$

where θ is the angle of incidence of the beams with respect to the normal of the plane, and k is the wavelength defined by $k = 2\pi/\lambda$.

In the case of nonlinear media, an interference pattern can be generated through an induced birefringence. The induced birefringence consists of the modulation of polarization and simultaneous intensity. When two optical beams are coupled in a nonlinear medium, interference occurs that induces a birefringence in the medium.

Thus, the intensity and polarization modulation of the light results from the interference of two optical waves derived from the two-wave mixing method. Also, the induced birefringence can be represented in the form of a diffraction grating. The interaction of the beams defines the grating that modifies the refractive index in a nonlinear medium that self-diffracts the waves. Then, the diffraction grating can be induced in a nonlinear medium through the spatial modulation of the polarization of the light generated by the optical wave mixing technique. Normally, the light diffracted by the grating generated by birefringence has a different polarization than the polarization of the incident light. In the case of a thin optical medium, there is a multitude of refracted waves. The simulations of the diffraction grating allow knowing the behavior of the vector amplitude of the light when leaving the nonlinear medium, resulting in the amplitude of the incident light and the transmittance of the medium by this way.

The nonlinear optical phenomena are associated with the wave properties modification of the light derived from the interaction of the electromagnetic wave with a material. Nonlinear optics describes the light behavior when the dielectric polarization component responds to the nonlinearity of the light electric field. Therefore, the interactions of electromagnetic fields produce alterations in phase, frequency, amplitude, or other propagation characteristics of incident waves. In the interaction of the light with a linear medium, the electromagnetic wave induces a polarization P_{NL} that depends linearly on the electric field strength E. On the other hand, in nonlinear optics, the polarization response can be described with Taylor's series in terms of the electric field as [22]

$$P_{NL} = \left[\chi^{(1)} E + \chi^{(2)} E^2 + \chi^{(3)} E^3 + \ldots \right] \tag{1.42}$$

where $\chi^{(1)}$ is the linear behavior of susceptibility and creates linear phenomena, $\chi^{(2)}$ and $\chi^{(3)}$ are the second- and third-order nonlinear susceptibility and creates second- and third-order phenomena. The third-order optical nonlinearities in materials can be represented by a $\chi^{(3)}$ tensor.

The components of the third-order nonlinear optical susceptibility tensor $\chi^{(3)}$ can be described as

$$\chi^{(3)}_{1111} = \chi^{(3)}_{1122} + \chi^{(3)}_{1212} + \chi^{(3)}_{1221} \tag{1.43}$$

The analysis of self-diffraction due to the mixing of optical waves through the Raman–Nath conditions shows the differences in the polarization characteristics of diffracted light in media with different physical mechanisms of nonlinearity. Furthermore, it is shown that the interaction of the orthogonal polarizations of the beams causes self-diffraction in media with nonlinearities associated with the physical and chemical characteristics of the environment. The results of the analysis allow determining the nonlinear mechanisms due to the intensities of polarized self-diffraction.

It is observed that, through Bragg and Raman–Nath theories of self-diffraction, it is possible to address the argument related to phase mismatch. Moreover, it is possible to study the arbitrary polarization coming from the signal wave and the linear or circular polarization of the wave with greater intensity.

Through the coupling of out-of-phase waves, it is possible to adapt the phase through a change in the refractive index of the medium. The amplification necessary for this change is obtained by the strong intensity beam induced in a Kerr medium. Furthermore, in metallic anisotropic media, the study of susceptibility tensors helps to understand the physical mechanisms that modify the optical response of materials. So, it is possible to define a form of control in the nonlinear optical response of metallic materials.

The third-order nonlinear optical effects are considered of utmost importance in the field of nonlinear optics because, although the second-order optical nonlinearities are of higher intensities, the second-order optical nonlinearities are only exhibited through noncentrosymmetric materials. At the time, in the case of centrosymmetric materials, being amorphous they are very likely to exhibit third-order optical non-linearities. Furthermore, third-order effects have the quality of acquiring phase coupling characteristics.

Among the nonlinear mechanisms that can prompt a third-order nonlinear optical phenomenon are the optical Kerr effect and the multiwave mixing. The multiwave mixing with short and ultrashort pulses is of relevance since it is derived from the comparison of the light intensities of the interacting polarized waves. The incidence of these waves causes a self-diffraction. Furthermore, in the nanosecond regime, thermal mechanisms responsible for optical nonlinearity, refractive index, and changes in the physical and chemical properties of materials are generated. In addition, the multiwave mixing is very functional in the use of wavelength sources, both infrared and ultraviolet; also, derived from the interaction of waves, existing energy exchanges are possible due to the frequency of the waves.

Depending on the energy interactions generated, an induced refractive index by the optical Kerr effect can be strong. Through the optical Kerr effect, it is possible to induce different mechanisms of optical nonlinearity in the material, which in turn generates the modification in the propagation mode of the beams. Thus, the optical Kerr effect presented in the materials modifies the nonlinear refractive index. This indicates that the refractive index of the material depends on the intensity of the incident light.

Therefore, in terms of the refractive index as a characteristic in the nonlinearities of a material, it is necessary to consider the influence of the polarization vector in terms of third-order susceptibility. In addition, the angle of propagation of mono-chromatic waves can be also an influence in optically induced nonlinear effects.

Mathematically, the response of the variation of the refractive index exhibited by a material at a certain frequency is characteristic of the nonzero third-order suscep-tibility tensor elements. On the other hand, in terms of physics, there are several physical mechanisms linked to the third-order susceptibility tensor. These mecha-nisms give rise to a change in the refractive index of the materials, and these can be optical, electronic, thermal, and molecular, among others.

1.5   Optical Nonlinearities in Reference Material for System Calibration

The physical phenomena related to third-order nonlinear optical effects are closely related to the proportional effects of high irradiance on materials. Therefore, nonlinear optical mechanisms are associated with molecular redistribution from optical absorption and the reorientation of anisotropic material molecules. In the case of materials with anisotropic characteristics, the atoms or molecules acquire a dipole moment induced under the action of an external field. Its origin is found in the deformation of the electronic distribution caused by the field. The polarizability induced by nonlinear optical interactions explains the fluctuation of the local field in each molecule due to the induced torque of the surrounding molecules. The polarized waves incident through the field will experience a local increase in the refractive index. Higher-order interactions among molecular-induced dipoles have an important influence on optical spectra of materials with strongly polarizable molecules, such as carbon disulfide (CS_2) [23].

There are experimental techniques used to determine third-order nonlinear optical properties of materials. Such experiments as z-scan, multiwave mixing, and optical Kerr effect measure transient nonlinear optical response can be characterized by the third-order optical susceptibility $\chi^{(3)}$.

The two-wave mixing method, as seen in Fig. 1.13, is used to measure the nonlinearity coefficients in the samples. This method allows identifying physical and thermal mechanisms that produce the third-order nonlinearities. Furthermore, with the mixing method, the energy transfer mechanism between light beams is manifested. In contradistinction to the z-scan technique, this cannot discriminate between the physical phenomena that cause the change in the nonlinear property of the medium being measured.

In measurement methods, CS_2 is generally chosen as the reference. Due to its simple chemical structure with double bonds and anisotropic properties, CS_2 plays a particular role in nonlinear optics experiments as a standard reference material. This material functions as an easily accessible calibration sample. Also, it does not exhibit absorption bands for popular laser wavelengths, resulting in the regularity of its nonlinear coefficients.

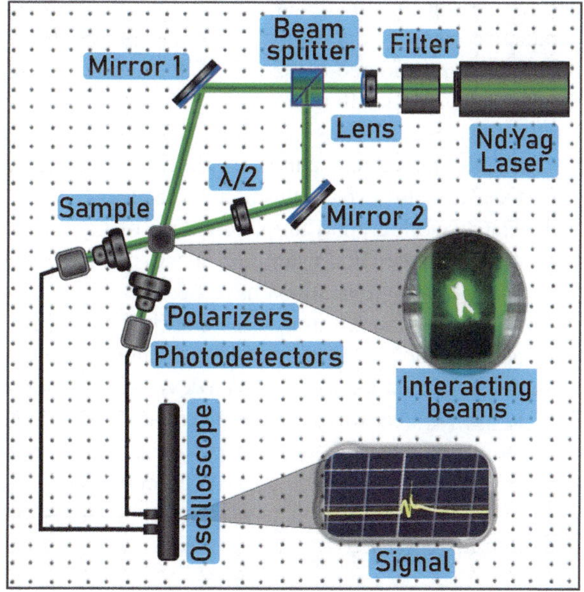

Fig. 1.13 Diagram of a two-wave mixing setup

Based on the values of these susceptibility elements, it is possible to show contributions to $\chi^{(3)}$. The reference value of CS_2 should be of the order of $\times 10^{-14}$ esu in a regime fs [24], in addition to coefficient of the intensity-dependent refractive index: $n_2 = 3.2 \times 10^{-14}$ cm^2/W [22]. In addition, an absorption value of two-photon coefficient $\beta = 8.7 \times 10^{-11}$ cm/W is highlighted for CS_2 at a wavelength of 532 nm. Measurements at the 532 nm wavelength exhibited much higher values of the two-photon absorption parameter [25]. The CS_2 demonstrates the constancy of its third-order nonlinearity numbers.

The study of nonlinear optical properties in materials has had a growing interest due to the technological applications developed and with potential to be developed. That is why, throughout this work, the importance of the study of nanomaterials has been defined, in addition to the physicochemical variables that govern the responses to the higher-order nonlinear optical effects of materials. In addition, the mechanical, optical, and electronic characteristics exhibited by materials susceptible to third-order optical nonlinearities have been highlighted. That is why it is important to point out the functionality of the materials studied through the synthesis that allows visualizing the field of potential applications. It has also been described that the nonlinear effects of materials originate through electronic interactions within matter because of the irradiation of high-energy beams.

On the other hand, organic materials and inorganic semiconductors are also good candidates for the development of nonlinear optical systems applications. Furthermore, organic materials are of great interest due to their low cost and ease of integration in device manufacturing. The plasmonic properties of the metallic nanoparticles allow increasing the emission and the control of the optical response due to the excitation of the electrons of the metallic structures. Furthermore, plasmonic behavior derived from decorations with metallic nanoparticles can function as an oscillator for frequency modulation. Also, gold nanoflowers have a high conductive capacity due to their size and morphology because of which potential materials are considered in the design and application of energy storage devices. On the other hand, metallic nanostars are considered ideal for biological detection applications because they have high dielectric sensitivity, as well as near-infrared resonances. This defines nanostars as ideal structures for the development of substrates sensitive to surface plasmon resonances.

Research work around the coefficients of nonlinear optical responses is also of utmost importance. For example, due to the nonlinear absorption coefficient of materials, various applications have been developed, including optical limiters and switches. Furthermore, through the coefficients, biomedical applications have been developed for the acquisition of biomedical images through high-speed multiphoton and confocal excitation fluorescence devices. Furthermore, the self-diffraction effect derived from optical wave coupling is related to nonlinearities derived from thermo-optic mechanisms and is useful for applications in optical limiting devices. It can then be concluded that, for the development of technologies derived from the study of third-order nonlinear optical responses, materials that present responses of large magnitudes are needed.

Research work around the coefficients of nonlinear optical responses is also of utmost importance. For example, due to the nonlinear absorption coefficient of materials, various applications have been developed, including optical limiters and switches. Furthermore, through the coefficients, biomedical applications have been developed for the acquisition of biomedical images through high-speed multiphoton and confocal excitation fluorescence devices. Furthermore, the self-diffraction effect derived from optical wave coupling is related to nonlinearities derived from thermo-optic mechanisms and is useful for applications in optical limiting devices. It can then be concluded that, for the development of technologies derived from the study of third-order nonlinear optical responses, materials that present responses of large magnitudes are needed.

References

1. Shen, Y. R. (1984). The principles of nonlinear optics.
2. Panoiu, N. C., Sha, W. E., Lei, D. Y., & Li, G. C. (2018). Nonlinear optics in plasmonic nanostructures. *Journal of Optics, 20*(8), 083001.
3. Ciriolo, A. G., Negro, M., Devetta, M., Cinquanta, E., Faccialà, D., Pusala, A., & Vozzi, C. (2017). Optical parametric amplification techniques for the generation of high-energy few-optical-cycles IR pulses for strong field applications. *Applied Sciences, 7*(3), 265.
4. Carrion, L., & Girardeau-Montaut, J. P. (2000). Development of a simple model for optical parametric generation. *JOSA B, 17*(1), 78–83.
5. Zhang, J. Y., Huang, J. Y., & Shen, Y. R. (2019). *Optical parametric generation and amplification*. Routledge.
6. Cho, M. J., Choi, D. H., Sullivan, P. A., Akelaitis, A. J., & Dalton, L. R. (2008). Recent progress in second-order nonlinear optical polymers and dendrimers. *Progress in Polymer Science, 33*(11), 1013–1058.
7. De Araújo, C. B., Gomes, A. S., & Boudebs, G. (2016). Techniques for nonlinear optical characterization of materials: A review. *Reports on Progress in Physics, 79*(3), 036401.
8. Chng, T. L., Orel, I. S., Starikovskaia, S. M., & Adamovich, I. V. (2019). Electric field induced second harmonic (E-FISH) generation for characterization of fast ionization wave discharges at moderate and low pressures. *Plasma Sources Science and Technology, 28*(4), 045004.
9. Collins, J. T., Rusimova, K. R., Hooper, D. C., Jeong, H. H., Ohnoutek, L., Pradaux-Caggiano, F., & Valev, V. K. (2019). First observation of optical activity in hyper-Rayleigh scattering. *Physical Review X, 9*(1), 011024.
10. Kajzar, F., Lee, K. S., & Jen, A. K. Y. (2003). Polymeric materials and their orientation techniques for second-order nonlinear optics. *Polymers for Photonics Applications, II*, 1–85.
11. Kleinman, D. A. (1962). Theory of second harmonic generation of light. *Physical Review, 128*(4), 1761.
12. Veithen, M., Gonze, X., & Ghosez, P. (2005). Nonlinear optical susceptibilities, Raman efficiencies, and electro-optic tensors from first-principles density functional perturbation theory. *Physical Review B, 71*(12), 125107.
13. Fuentes-Hernandez, C., Ramos-Ortiz, G., Tseng, S. Y., Gaj, M. P., & Kippelen, B. (2009). Third-harmonic generation and its applications in optical image processing. *Journal of Materials Chemistry, 19*(40), 7394–7401.
14. Thiel, C. W. (2008). *Four-wave mixing and its applications*. Faculty of Washington.
15. Simkhovich, B., & Bartal, G. (2014). Plasmon-enhanced four-wave mixing for superresolution applications. *Physical Review Letters, 112*(5), 056802.

16. Kovsh, D. I., Yang, S., Hagan, D. J., & Van Stryland, E. W. (1999). Nonlinear optical beam propagation for optical limiting. *Applied Optics, 38*(24), 5168–5180.
17. Smith, N. A., & Meech, S. R. (2002). Optically-heterodyne-detected optical Kerr effect (OHD-OKE): Applications in condensed phase dynamics. *International Reviews in Physical Chemistry, 21*(1), 75–100.
18. Torres-Torres, C., & Khomenko, A. V. (2005). Autodifracción vectorial de dos ondas degeneradas en medios con efecto Kerr óptico. *Revista mexicana de física, 51*(2), 162–167.
19. Castro-Chacón, J. H., Khomenko, A. V., & Rangel-Rojo, R. (2009). Phase matched vectorial three-wave mixing in isotropic Kerr media. *Optics Communications, 282*(7), 1422–1426.
20. Qasymeh, M., Cada, M., & Ponomarenko, S. A. (2008). Quadratic electro-optic Kerr effect: Applications to photonic devices. *IEEE Journal of Quantum Electronics, 44*(8), 740–746.
21. Reyes-Esqueda, J. A., Rodríguez-Iglesias, V., Silva-Pereyra, H. G., Torres-Torres, C., Santiago-Ramírez, A. L., Cheang-Wong, J. C., Crespo-Sosa, A., Fernández-Rodríguez, L., López-Suarez, A., & Oliver, A. (2009). Anisotropic linear and nonlinear optical properties from anisotropy-controlled metallic nanocomposites. *Optics Express, 17*(15), 12849–12868.
22. Boyd, R. W. (2020). *Nonlinear optics*. Academic press.
23. Kiyohara, K., Kamada, K., & Ohta, K. (2000). Orientational and collision-induced contribution to third-order nonlinear optical response of liquid CS_2. *The Journal of Chemical Physics, 112*(14), 6338–6348.
24. Yan, X. Q., Zhang, X. L., Shi, S., Liu, Z. B., & Tian, J. G. (2011). Third-order nonlinear susceptibility tensor elements of CS_2 at femtosecond time scale. *Optics Express, 19*(6), 5559–5564.
25. Ganeev, R. A., Ryasnyanskiĭ, A. I., & Kuroda, H. (2006). Nonlinear optical characteristics of carbon disulfide. *Optics and Spectroscopy, 100*(1), 108–118.

Chapter 2
Theoretical Models for Describing an Enhancement in Nanoscale Optical Nonlinearities for Applications Design

2.1 Introduction

The applications related to the control of the optical response have advanced in parallel with the understanding of the nature of the behavior of light in a medium.

Thus, the design, manufacture, and study of optical properties of nano-dimensioned systems with potential applications are constantly growing. Therefore, the study of optical systems made up of various nanostructures, among which are metals and semiconductor materials with different dimensions and geometries, has been of utmost importance, such as nanoparticles, wires, thin layers, and some other complex geometries.

Furthermore, the optical properties of these systems are in the domain of the characteristics of the materials that make them up, in addition to shape, physical–chemical environment, and even the interaction with other molecules and nanostructures. On the other hand, the nonlinear optical response associated with the various optical systems can be influenced by external mechanisms related to the nature of the nanostructure or independent physical mechanisms related to the environment or the characterization methods of nanostructured optical systems. Thus, the localization of light in nanostructured systems has made it possible to overcome the limits of conventional optics, increasing the possibilities of controlling various optoelectronic processes.

In conclusion, understanding the interaction between light and matter at the nanoscale opens a gap for the generation, control, and manipulation of light beams in tiny spaces. With the above, it is possible to establish a large number of technological possibilities.

© The Author(s), under exclusive license to Springer Nature Switzerland AG 2022

C. Torres-Torres, G. García-Beltrán, *Optical Nonlinearities in Nanostructured Systems*, Springer Tracts in Modern Physics 287, https://doi.org/10.1007/978-3-031-10824-2_2

2.2 Theoretical Models for Describing the Influence of Quantum Confinement in Optical Nonlinearities

A quantum confinement system is also called a low-dimensional system. This system is characterized because in it the wearers have the freedom to move in two, one, or none of the three spatial directions or dimensions. The dimensions are also known as the quantum scale. Furthermore, the direction of confinement in spatial dimensions defines the order of the De Broglie wavelength of the carriers. There are various experimental methods to obtain confined quantum systems. Most of these experimental methods are related to the application of some electromagnetic field or through the methods of synthesis of the materials.

Nanostructured materials are defined as real confined quantum systems. The above can be referred to due to the synthesis of nanostructures through semiconductors, metals, dielectrics, and magnets. Because of these materials, the state of the carriers is affected by the reduction to a quantum scale of the spatial dimensions.

The emergence of nonlinear optical effects occurs through the application of an optical field of greater magnitude to the atomic fields of the material system. Thus, the effects due to size become important in optical processes. When the dimensions of a system are equal to or are smaller than the natural length scale, it leads to the effects related to its quantum mechanics. On the other hand, the interactions related to the forbidden band induced through incident light originate from the transitions between bands. These transitions originate at the atomic level from the atoms that make up matter. In addition, the pairs of electrons and holes that participate in the absorption and emission processes need to be delocalized. Similarly, electrons and holes are strongly related through a long-range Coulomb-derived interaction. The interactions between pairs of electron–holes result in atomic and collective behavior that also gives rise to complex properties. Furthermore, the pairs of electron–holes form Wannier excitons, whose Bohr radius α_0 is very useful for determining the necessary length for the origin of optical effects in a semiconductor structure [1].

Also, electron–hole pairs generated within quasi-two-dimensional active layers are derived from the excitation of the laser beam. So, these layers are responsible for the quantum confinement effect. Also, quantum effects are important in enhancing excitonic effects.

The impact of quantum confinement can be understood through the spectral characteristics of the Bloch states of low wave numbers k. Thus, quantum confinement can be perceived in the holes and pairs of electrons derived from the excitation of a beam of light in semiconductors, as well as surface impurities. Derived from the above, the scale modulations less than $1/k$ change the energies smoothly and the following equation can be established [2]:

$$\varepsilon_i(k) = \pm\left(\frac{E_g}{2} + \frac{\hbar^2 k^2}{2m^* i}\right) \tag{2.1}$$

where ε is the dielectric constant bulk, m_e^* is the effective mass of the electron, m_h^* refers to the effective mass of the hole, $i = $ e, h refers to the conduction states of electrons and valence holes, respectively, and $m > m_h^* > m_e^*$. In addition, the wave functions are established through

$$\varphi_i(r) = F_i(r)u_i(r) \tag{2.2}$$

where $F_i(r)$ is a wave envelope that varies through cells, and $u_i(r)$ refers to the periodic cell of the Bloch wave functions with close to $k = 0$. It is essential in this case to determine the envelop $F_i(r)$. Then, the condition of kinetic and potential energy for an electron–hole pair delimits the extension of this shell. The previous condition is fulfilled when the spacing between the radius of the energy level is delimited by an intrinsic property of the volume of the semiconductor associated to ae and ah. Radii have a relationship of $a_e = h^2\varepsilon/m_e^* \, e^2$ for electrons, and $a_h = h^2\varepsilon/m_h^* e^2$ is associated with holes.

Then, if the electron–hole pair is formed in a confined region of radius smaller than ae or ah, there is an interaction of the electron–hole that is partially or totally suppressed in a favorable way to the kinetic energy. Consequently, the electron and hole move as free particles with an effective mass of m^*_i in a quantum well of spherical shape and infinite height. So, the states of the quantum well are calculated differently, such that the oscillator intensities for the transitions between energy levels undergo a redistribution.

Then, derived from the above, it is possible to distinguish three main regimens of quantum confinement, which, as a principle, correspond to the relative position of the radius a with respect to a_e and a_h. For the regimens of quantum confinement, it must be considered that $m_e^* < m_h^*$ and by then $a_e > a_h$.

For strong quantum confinement, it is stated that $a < a_h < a_e$. Also, it is established that the interaction between the electron–hole is suppressed, and the movements of electrons and holes are uncoupled. Thus, the motions of electrons and holes are quantized separately in a spherical quantum well of radius and infinite potential height. So, the energy levels are indicated by the quantum numbers nl. Also, only transitions between levels with the same quantum numbers are possible. In particular, the lowest allowed transition is expressed as

$$\hbar\Omega_0 = E_g + \hbar^2\pi/2\mu a^2 \tag{2.3}$$

where $\mu = m_h^* \, m_e^*/(m_h^* + m_e^*)$. It is then observed that the oscillator force is focused on transitions and the nonlinear optical properties improve.

The intermediate quantum confinement establishes the condition $a_h < a_e < a$; then, the movement of the electron is quantized through the movement of the hole. Furthermore, the state of the electron is calculated through an adiabatic approximation. Therefore, each electronic transition becomes a series of closely spaced lines with an asymmetric envelope.

Finally, the intermediate quantum confinement establishes that $a_\mathrm{h} < a_\mathrm{e} \le a$. Also, the movements of electrons and holes are synthesized through their coupling and volume characteristics. In particular, the movement of bound exciton states is quantified within the crystallite such that the displacement of the lowest transition of the exciton is given regarding its position in the mass in an amount that can be represented as

$$\Delta = \hbar^2 \pi^2 / 2Ma^2 \tag{2.4}$$

where $M = m_\mathrm{e}{}^* + m_\mathrm{h}{}^*$.

Moreover, through the quantum confinement regimens, it is possible to establish the modification of the third-order nonlinear optical susceptibility $\chi^{(3)}$ where the Hamiltonian H of the Coulomb system with undisturbed electron–holes in two-band approximation is denoted such that the time-dependent total Hamiltonian is expressed as [3]

$$H_t = H - PE \cos\left(\omega t\right) \tag{2.5}$$

where E is the applied electric field, and P is associated with the polarization between bands. On the other hand, $t = 0$ is considered. The above establishes that the system is in its fundamental state, without electrons or holes. Thus, the third-order optical sustainability due to quantum confining interactions is established through

$$\chi^{(3)} = \begin{aligned} &-\frac{iE^2}{4V} \left\{ \frac{4A^2}{(i\omega_1 + \gamma)(\omega_1^2 + \gamma^2)} - \sum_\sigma B_\sigma^2 \left[\frac{2}{[i(\omega_{2\sigma} - \omega_1) + \gamma](\omega_1^2 + \gamma^2)} \right. \right. \\ &\left. \left. + \frac{1}{(i\omega_1 + \gamma)(i\omega_{2\sigma} + \gamma)} \left[\frac{1}{i(\omega_{2\sigma} - \omega_1) + \gamma} - \frac{1}{i\omega_1 + \gamma} \right] \right] \right\} \end{aligned} \tag{2.6}$$

The third-order optical susceptibility $\chi^{(3)}$ generally exhibits a resonant improvement due to the proximity of a resonance. Thus, the third-order nonlinear optical response can be considered sensitive to the characteristics of confined quantum resonances. The influence of quantum confinement effects on third-order optical susceptibility can be mediated through an optical wave mixing method such that, as field strengths increase, the effects higher than those of the third order are involved and the optical absorption coefficient and Kerr become dependent on the intensity close to the transition frequency of a two-level system.

On the other hand, the effects of quantum confinement associated with semiconductor nanostructures lead to interventions on nonlinear optical properties. Therefore, the nonlinearities influenced by the quantum confinement of nanostructures have received great attention due to the technological advantages that could be developed through it. Moreover, the study of quantum confinement effects at the nanoscale has presented a series of complications due to the lack of manufacturing techniques that allow obtaining nanocrystals with narrow distributions and semiconductor–dielectric interfaces. Furthermore, the size of the semiconductor crystals directly affects the intrinsic effects of quantum confinement to the extent of hiding them since an extrinsic broadening is introduced that can dominate the intrinsic broadening in the confined quantum nanostructure. For this reason, nonlinear optical techniques allow the extraction of relevant information on the effects of quantum confinement and on the potential use of these materials in nonlinear optical devices. One of the techniques used to evaluate the importance of confined quantum scaling mechanisms is the time-resolved spectral hole burning on nanosecond, picosecond, and subpicosecond time scales. Through this nonlinear technique, it has been established that nonlinear absorption measurements show that, at low-temperature nonhomogeneous broadening dominates, at room temperature the dominant broadening mechanism is intrinsic or homogeneous and the ls–ls transition saturates uniformly. This saturation establishes the nonlinear mechanism. So, when moving from a weak confinement where the oscillator force is distributed over a wide band to a confined case where this oscillator force is now condensed into discrete lines, an improvement in nonlinearity is expected.

To describe the properties of semiconductor nanostructures, a strong confinement regime is assumed. Also, optical nonlinearities can be due to various mechanisms. The first mechanism is saturation of the ls–ls transition. When the 1 s–1 s peak is sufficiently isolated, each particle behaves as a two-level system that can be bleached. In the second mechanism, two electron–hole pairs can be excited per particle and the transition between the state of one pair and the state of two pairs leads to induced absorption. Due to the Coulomb interaction, this last transition is shifted in frequency compared to ω_{1s}. The third mechanism assumes that the photoexcited carriers are trapped on the surface of the particle and the static electric field that they create modifies the absorption spectrum. So, considering the first mechanism, it is established that the Kerr susceptibility is obtained through a two-level system such that, for particles of radius a, we obtain

$$\chi_a^{(3)} = (\omega, -\omega, \omega) = -N\frac{d^4}{\hbar^3}T_1 T_2^2 \frac{1}{\delta^a - i} \cdot \frac{1}{(\delta^a)^2 + 1} \tag{2.7}$$

On the other hand, if we consider confined particles for which the 1s–1s transition is isolated, then for the $\omega \simeq \omega_{1s}$ we obtain

$$\alpha^a(\omega) \simeq \alpha_{1s}^a(\omega) = \frac{\omega}{nc}4\pi N\frac{d^2}{\hbar}\frac{T_2}{(\delta^a)^2 + 1} \tag{2.8}$$

The volume fraction p is given by

$$p = N\frac{4\pi}{3}a^3 \tag{2.9}$$

So, for a given volume fraction, both α_a and $\chi_a^{(3)}$ would be inversely proportional to the volume of the particle and, therefore, would be improved by quantum confinement, such that [4]

$$\frac{\chi_a^{(3)}}{\alpha^a} \propto d^2 \frac{T_1 T_2}{\delta^a - i} \tag{2.10}$$

It is called exciton to pairs of electrons attached to holes in a semiconductor. Exciton polaritons are referred to hybrid semiconductor excitations. The excitations arise through particles when there is a strong coupling regime due to the light–matter interaction between photons and excitons, which also exceeds the dissipation rate of semiconductors. For the study of nonlinear optics, exciton–photon interactions have been considered because they are strong, while exciton–exciton interactions exhibit orders of magnitude smaller. Therefore, polaritons expand by orders of magnitude in response to the combination of high photonic coherence and optical nonlinearities such that, in the range of parameters, a variety of quantum phenomena can be maintained [5].

Moreover, excitonic effects are particularly important in two-dimensional nano-structured materials due to quantum confinement effects. That is why excitonic effects play an important role in determining the linear and nonlinear responses of two-dimensional nanostructured materials, particularly in the case of materials that exhibit a wide bandgap response. In the particular case of optical responses, excitons are manifested through a shift in the UV-Vis spectrum toward red [6].

It has also been defined that the size of the nanoparticles is a very important factor in the determination of the optical nonlinearities of the nanostructures. This is due to the fact that the size is directly related to the spatial distribution of the polarization charges on the surface of the nanoparticles. So, when there is a variation in the size of the nanostructures, there is also a change in factors such as plasmon bandwidth and surface plasmon resonance wavelength. Equally, third-order optical nonlinearities have a direct dependence on the size of the nanoparticles. Furthermore, the nano-structure size smaller than 10 nm facilitates the quantum confinement effect in nanoparticles. Thus, the optical nonlinearities of nanostructures also depend on quantum confinement effects in a specific size range [7].

Thus, the quantum confinement of the carriers leads, at discrete levels within the conduction and valence bands, to a shifted and structured absorption spectrum. Besides, the nonlinear optical response mechanism is modified. Through the close-ness to the first 1 s–1 s transition, each particle behaves as a two-level system whose saturation is responsible for the nonlinearity. However, the nonlinearities are limited by the width of the 1 s–1 s transition. Thus, through confined quantum systems, it is possible to associate the size and frequency dependence of the third-order optical susceptibility $\chi^{(3)}$ [8].

2.3 Theoretical Models for Describing the Influence of Plasmonic Phenomena in Optical Nonlinearities

As has been established, nonlinear optical effects are of importance for optical signal processing. Thus, plasmonic nanostructures are widely used to increase the efficiency of light–matter interactions. Also, plasmonic effects are related to a coupled state of photons and free-carrier coherent oscillations near a conductor–dielectric interface.

Plasmonic effects are referred to as plasmonic excitations, plasmon polaritons, or surface plasmons located on metallic nanoparticles. Plasmonic excitations establish confinement of the electromagnetic field, which results in enhancement of the local field. Moreover, plasmonic excitations lead to strong enhancements over nonlinear optical processes since they depend on the local field intensity. From the above, improvements are also generated near the metal–dielectric interface in the presence of plasmonic resonances. Therefore, materials in the nanoscale range establish an advantage because it is possible to establish a nonlinear response in sub-wavelength scales.

Then, it is possible to induce a nonlinear optical response through the field enhancement provided by the surface plasmons. Since the nonlinear response originates from the dynamics of free electrons that are not in equilibrium in the medium under the influence of the strong electromagnetic field derived from the irradiated light beam, it is then established that the nonlinear response of the metal is one of the strongest per unit length of interaction and one of the fastest.

It has been established that when the local electric field is enhanced, there may be an increase in the nonlinear susceptibility close to the resonances of a plasmonic structure. Furthermore, whether the electromagnetic resonances and field enhancement at the generated frequency of the nonlinear harmonic near field, as well as its radiation in the far field, are susceptible is exhibited. From the above, the effective nonlinear optical susceptibility is expressed as [9]

$$\chi^{(n)}_{\text{eff}} = E(n\omega)L(n\omega)L^n(\omega)E_0^n(\omega) \tag{2.11}$$

Thus, the local electromagnetic field can be enhanced through plasmonic nanostructures since they provide high local electromagnetic fields. Then, it is possible to establish that plasmonic nanostructures are good candidates for the enhancement of nonlinear optical effects [10]. Furthermore, the relationship between the phases and the fundamental and harmonic frequency modes is of importance for the generation of higher-order harmonics from metallic nanostructures.

On the other hand, Kerr nonlinearities are also associated since plasmonic excitations are sensitive to changes in the refractive index, in addition to having the advantage that it is not only sensitive in the metallic nanostructure, but also in the interface or in the surrounding material. The change in the nonlinear refractive index can be induced through the propagation of a signal beam in waveguides and the transmission of light, which in turn establishes the modification of plasmonic resonances.

Then, from third-order Kerr nonlinearities, the intensity-dependent refractive index can be modified, expressed as

$$n(I) = n_0 + \gamma \left| E(\omega_c) \right|^2 \qquad (2.12)$$

Equally, the absorption is expressed as

$$\alpha(I) = \alpha_0 + \beta \left| (\omega_c) \right|^2 \qquad (2.13)$$

where n_0 corresponds to the linear refractive index and α_0 to the absorption. Furthermore, $|E(\omega_c)|^2$ expresses the control light intensity, and n_2 sets the absorption coefficients. Thus, nonlinear refractive index and absorption can alter the phase of the light signal at a different frequency than the one that interacts with the nanostructure. This phenomenon is also called modulation. In addition, there is also the phenomenon of self-modulation that originates because the intensity of the light can influence itself considering the same optical effects. In addition, the field enhancements associated with plasmonic resonances provide the origin of Kerr optical nonlinearities from the reduction of the required control light intensities [11].

Plasmonic nanostructures exhibit good nonlinear optical response due to the dynamics of electrons in the inhomogeneous electromagnetic field. In addition, the optical nonlinearities associated with metallic nanostructures allow the design of applications in the field of ultrafast optical signal processing.

On the other hand, through geometric modifications, such as decreasing the size of the nanostructure, it is possible to observe additional contributions due to the inhomogeneous field and nonlocal effects. Also, the Kerr nonlinearities are determined through the heating of electrons through their redistribution of energy in a conduction band. In addition, they also depend on the energy of the light incident on the electron plasma. So, the nonlinearities have a dependence on the energy and duration of the pulse [12]. In addition, the nonlinearities derived from metallic nanostructures are associated in origin from several families of physical effects such as the nonlinear dynamics of free carrier plasma and interband electronic excitation, when the conducting field excites valence band electrons to drive them. Electronic excitation has a high efficiency associated with absorption. However, it has a spectral range limited to optical transitions between bands [13].

For metallic nanoparticles, extrinsic effects are mainly responsible for size-dependent optical phenomena. However, it is also possible to detect intrinsic size-dependent properties. Therefore, for the optical response of metallic nanoparticles, extrinsic effects are of great importance, while intrinsic properties are of secondary importance [14].

On the other hand, the optical response of metallic nanostructures can be described through Mie's theory. This theory consists of calculating the extinction coefficient of metallic nanostructures considering their complex dielectric function called ε_{NP}, embedded in a host that presents the dielectric function, which is also represented as ε_h. Besides, noble metal nanostructures have been considered to have a broad optical response due to the interaction of electrons with the optical field and some physical processes. The physical processes are referred to in-band transitions, which consist of transitions related to electrons in the conduction band, and on the other hand, interband transitions that refer to transitions of electrons from the valence to the conduction band [15].

On the other hand, it has been established that the main contribution to the optical absorption of metallic nanomaterials is due to the localized surface plasmon resonance, while for photon energies greater than $\sim$4 eV, the transitions between bands strongly influence the optical response. Furthermore, when the dimensions of the metallic nanostructures are much smaller than the wavelength λ of the beam, the light field is uniform throughout the particle and the collective oscillation of the electrons is described by a dipole polarizability represented by

$$\alpha = (1 + n)\varepsilon_0 V \frac{(\varepsilon_{NP} - \varepsilon_h)}{(\varepsilon_{NP} - n\varepsilon_h)} \tag{2.14}$$

where V is the volume of the metallic nanostructures, and η is a geometrical factor that depends on the shape of the metallic nanostructures, which in turn, $\eta = 2$, is established for spherical nanoparticles. Consequently, when the polarizability of the nanostructures becomes maximum, and the corresponding frequency is the localized surface plasmon resonance frequency, so, the frequency related to the localized surface plasmon called ω_{LSP} can be expressed as

$$\omega_{LSP} = \sqrt{\frac{Ne^2}{m_e\varepsilon_0 \, \text{Re} \left(\varepsilon^{ib} + n\varepsilon_h\right)} - \gamma^2} \tag{2.15}$$

Thus, it is observed that the wavelength of the localized surface plasmon resonance shifts toward the red when the particles are embedded in a host with larger Re (ε_h) such that the absorption response, described by Im (ε_{NP}), is mainly due to collisions between electrons and pons, in addition to scattering between electrons and the surface.

The localized surface plasmon has a hybrid electromagnetic wave and a surface charge that creates a field component that decays exponentially due to the distance between the nanostructures. Also, for interfaces between metal, localized surface plasmon excitation occurs in the ultraviolet to near-infrared region. Furthermore, in the case of metallic nanostructures with nonspherical geometries, they exhibit a localized surface plasmon resonance frequency lower than that exhibited by spherical metallic nanostructures.

So, in the case of spherical nanoparticles, the optical extinction coefficient in metallic nanostructures can be explained through Mie theory, with the following equation:

$$\alpha_{\text{ext}} = \frac{18\pi V_{\text{NP}}}{\lambda} \left[\varepsilon_h(\omega)\right]^{3/2} \frac{Im[\varepsilon_{\text{NP}}(\omega)]}{\left(\,Re\left[\varepsilon_{\text{NP}}(\omega)\right] + 2\varepsilon_{\text{h}}(\omega)\right)^2 + Im[\varepsilon_{\text{NP}}(\omega)]^2} \qquad (2.16)$$

where α_{ext} describes the contributions of light absorption and scattering.

It is observed, then, the size of the particle in relation to its amplitude defines the magnitude of the extinction coefficient. However, the position and bandwidth of the localized surface plasmon resonance have been shown to depend on the shape and size of the nanoparticles. Then, Mie theory was modified to consider the dependence of the size of the dielectric function of spherical metallic nanostructures, corresponding to ε_{NP} (ω, R), where, in addition, the radius of the nanoparticle corresponding to R is plug in the electron collision frequency, which refers to $\gamma = (\gamma_{\text{bulk}} + v_f/R)$. Thus, it is possible to observe that the localized surface plasmon resonance wavelength shifts toward the blue region as the particle size is reduced. Furthermore, the extinction spectra of the metallic nanostructures provide information about the studied metal.

On the other hand, there are important characteristics that can contribute to the spectrum profile, such as the full width at half the maximum, the frequency of localized surface plasmon resonance, and the amplitude of the curve. So, for different metals, the localized surface plasmon resonance frequency and the contribution of interband transitions in the optical absorption spectrum are different. Furthermore, through the spectrum and Mie theory it is possible to estimate the size of spherical nanoparticles. Furthermore, the size dispersion of metallic nanostructures in a set produces a broadening of the absorption band of localized surface plasmon and the deviation of the sphericity of the nanostructures produces a resonance shift of localized surface plasmon at larger wavelengths such that, for an individual nanostructure, the near electric field is the sum of the incident field, E_0, plus the field generated by the oscillation of electrons, so that, on the surface of the nanostructure, at the dipole limit, the near field is expressed as

$$E_{\text{NP}} = \frac{(1 + n)\varepsilon_{\text{h}}}{(\varepsilon_{\text{NP}} + n\varepsilon_{\text{h}})} E_0 \qquad (2.17)$$

2.4 Theoretical Models for Describing the Influence of Fano Resonances in Optical Nonlinearities

A resonance can be described by a harmonic oscillator with periodic forcing. When the frequency of the driving force is close to the frequency of the oscillator, the amplitude of the oscillator increases to a maximum value. On the other hand, there is also anti-resonance, in which its response is suppressed in the event of some resonance condition. Moreover, there are resonances in which there are two weakly coupled harmonic oscillators, where one of them is driven by a periodic force, such that both resonances are close to the frequencies of the oscillators, such that one of the resonances is characterized by a symmetric profile, and the second resonance is characterized by an asymmetric profile, which presents a total suppression of the amplitude of the forced oscillator at the frequency of the second oscillator. Similarly, the amplitude of the first oscillator becomes zero as a result of destructive interference from the oscillations of the external force and the second oscillator. So, the phase of the forced oscillator exhibits a jump in resonance, resulting in that under resonance the oscillator is in phase with the driving force and out of phase above resonance.

Fano resonances are characterized due to the interaction between a discrete state and a continuous state, and through the said interaction an asymmetric line shape is established, as described in the formula established by Ugo Fano. Such asymmetry is due to the spectral proximity between constructive and destructive interferences, resulting in spectral response characteristics [16].

Besides, Fano resonances have been of importance for physical phenomena such as absorption, Mie scattering, and Bragg scattering, in addition to the autoionization of atoms. Equally, the Fano resonances are essential in the treatment of the plasmon resonance width that generally has a Lorentzian shape and the near field. In addition, the Fano resonances result from the interaction between the modes of the system that derives in dark and bright modes. The bright mode originates due to a large dipole moment such that it is excited by plane-wave illumination from far-field scattering. The dark mode has a lower dipole moment and does not interact with the plane wave or scatter by the far field. The dark mode is excited via the near field, so the plane wave excites the bright mode, then excites the dark mode, and furthermore interferes with the scattering of the bright mode itself. Then, the bright mode response can be attenuated or enhanced based on the phase difference. The attenuated or enhanced characteristics of the Fano resonances allow the control of the optical properties of nanostructures and nonlinear optical systems [17].

So, to begin to understand Fano resonances, we can think of the case of two coupled oscillators at the resonance of the second oscillator, there are two driving forces acting on the first oscillator, which are out of phase and cancel each other such that the resonant destructive interference highlights the Fano resonances of others. In other words, Fano resonance occurs when the bright mode in an electromagnetic wave constructively and destructively interferes with the discrete dark mode. The said phenomenon is observed through the angle-dependent reflectivity of a Fano resonance that exhibits an asymmetric line-shaped profile.

So, Fano resonances can be represented by using a coupled-driven two oscillator model, which, in turn, can be described as

$$
\begin{pmatrix} \omega_1 - \omega - i\gamma_1 & g \\ g & \omega_2 - \omega - i\gamma_2 \end{pmatrix} \begin{pmatrix} x_1 \\ x_2 \end{pmatrix} = i \begin{pmatrix} f_1 \\ f_2 \end{pmatrix} \tag{2.18}
$$

where x_1 and x_2 represent the amplitudes of the oscillator, ω_1 and ω_2 refer to the resonance frequencies, γ_1 and γ_2 are the damping coefficients, f_1 and f_2 establish the external forces with the excitation frequency ω, and g represents the coupling constant that describes the interaction between the oscillators.

On the other hand, damping can be ohmic losses or leakage losses due to coupling to other modes such that the interaction between the oscillators generates a change in their complex frequencies. If the first oscillator behaves as a quantum emitter and the second oscillator corresponds to a photon mode of a cavity, then, the imaginary part of $\Delta\omega_1$ describes the modification of the rate of spontaneous emission in the cavity, also associated with the Purcell effect. Also, the real part of $\Delta\omega_1$ is the radiative correction in frequency associated with the Lamb shift.

On the other hand, Fano resonances can be represented through a model of two oscillators in the weak coupling regime when only one of the oscillators is forced. In other words, the value of the first external force is different from zero, and the second external force is equal to zero such that the amplitude of the first oscillator in the spectral zone of the resonance of the second oscillator can be represented as [18]

$$|x_1(\Omega)|^2 \approx |f_1{}^2|\frac{\gamma_1{}^2}{(\omega_1-\omega_2)^2+\gamma_1{}^2}\frac{(\Omega+q)^2}{(\Omega^2+1)} \tag{2.19}$$

where Ω refers to the dimensionless frequency and is represented as

$$\Omega = \left[\omega-\omega_2+\left(\frac{g^2}{\gamma_1}\right)\frac{(\omega_1-\omega_2)}{1+\gamma_1}\right]\frac{\gamma_1(1+q^2)}{g^2} \tag{2.20}$$

Also, q determines the spectral shape and depends on the spectral detuning of the oscillators ω_2–ω_1.

Another important point to note is that nonlinear optical processes can be improved and eliminated through constructive and destructive interferences corresponding to frequency conversion. Thus, the interference effects derived from Fano resonances in the optical nonlinearities can be exhibited by means of coupling to a classical medium such as a quantum one.

From another point of view, some effects favor the improvement of nonlinear optical processes, such that it is possible to determine the improvements and losses in nonlinear optical effects due to interference derived from Fano resonances. On the other hand, it has been established that there is an influence of Fano resonances on the nonlinear optical response of nanostructures with plasmonic properties. So, improvements can be established from the control of light generation and nonlinear effects at nanometric scales. Fano resonances have also been shown to be important for the design of plasmonic sensors based on second-order nonlinearities such that, from the second harmonic generation, the observation of multipolar modes with useful resonances in detection applications is possible.

2.4.1 Influence of Fano Resonances in the Localized Surface Plasmon

Of the wide range of existing nanostructures, plasmonic nanostructures are of high importance in the field of nanophotonics because they have characteristics with the ability to improve the intensity of the electromagnetic field in small volumes. The improvement of the electromagnetic field is possible to control through Fano resonances. Also, the Fano resonances are exhibited through multiple plasmonic nanostructures with complex geometries.

Plasmonic phenomena have been characterized by the relationship with the modification of responses of nanostructured light systems. Thus, Fano resonances have been studied in combination with plasmonic characteristics because they can alter a resonance in a very small frequency range. Also, Fano resonances have established a connection with plasmonics through enhanced surface infrared absorption in which a molecular absorption line interacts with localized surface plasmon resonances of metallic nanostructures. Moreover, the Fano resonances result from symmetrical breaks that can be induced through specular charges on the substrate or through cavities. Fano resonances are extremely important for plasmon-emitting systems.

Through metallic nanostructures, it is possible to observe Fano resonances because, through transverse magnetic polarization, the incident light can excite a plasmon of particles through metallic nanoparticles. This can destructively interfere with the resonance of another element such as a waveguide. Thus, the light incident to the waveguide destructively interferes with the re-emitted light from the particle plasmon. Then, when modifying the incident excitation angle, the spectral dispersion shows an anticrossing behavior [19].

The excitation of a plane wave in a metallic nanostructure drives the localized surface plasmon mode such that its frequency coincides and involves an interaction of the incident wave with the polarization field of the plasmon. This interaction is expressed as [20].

$$H_{\mathrm{LSP}} = \left(E_0 \int d^3(r)\epsilon(r)E_1(r).\widehat{x}e^{ikz} \right)\widehat{a}_1 e^{-i\omega t} + H.c. \tag{2.21}$$

where $E_1(r)$ corresponds to the amplitude of the excitation field. Different metallic nanostructures have established important behaviors related to Fano resonances, which indicates that these models provide very useful information to understand Fano resonances in nonlinear optical regimens. These results show the importance of the design of applications involving plasmonic sensors based on second-order nonlinear optical processes [21].

2.4.2 Influence of Fano Resonances in Second Harmonic Generation

The second harmonic generation depends significantly on the material symmetry of the studied nanostructure. In the same way, the second harmonic generation of centrosymmetric materials is possible through surface inversion symmetry breaking. For the technical second harmonic generation, the presence of nonlinear bias currents oscillating at the second harmonic frequency, on the surface of the nanostructure, is necessary, as well as an efficient dispersion of the second harmonic generation signal in the far field. However, the efficiency of second harmonic generation is limited to two fundamental classes of losses suffered by electromagnetic waves such as optical dispersion and heat generation losses. Therefore, to increase the generation of the second harmonic in the far field, it is necessary to decrease losses due to radiation in the wavelength. Similarly, the near field at the fundamental wavelength is increased relative to the second harmonic wavelength. Since second harmonic generation performance increases with the square of the field strength, a higher near field is necessary to increase the near field at the second harmonic. On the other hand, the resonance width of a plasmon indicates the total losses, radiative and nonradiative, through the nanostructure. That is why low losses are necessary to adjust the plasmon resonance width by modifying the structures.

On the other hand, localized plasmon modes are responsible for the second harmonic generation process in metallic nanostructures. The above, due to two plasmons that oscillate with frequency ω, combine to produce a single plasmon that oscillates with frequency 2ω. This can be described through a Hamiltonian expressed as

$$\widehat{H}_{\text{SHG}} = \left(\int d^3 r E_2^*(r) E_1^2(r) \chi_2(r) \right) \widehat{a}_2^\dagger \widehat{a}_1 \widehat{a}_1 + H.c. \tag{2.22}$$

where $\chi^2(r)$ is the second-order polarization, and it is also expressed to determine the strength of the second harmonic generation $\chi^{(2)} = d_3(r) \, E_2^*(r) \, E_2(r) \, \chi^2(r)$. Thus, second harmonic generation has been shown to be a remarkable technique for observing higher multipolar modes with narrow resonances [18].

2.4.3 Influence of Fano Resonances in Four-Wave Mixing

Fano resonance interferences have been associated with discrete single-state systems coupled to a continuous state such that it is possible to identify the Fano resonance interference if the strength of the oscillator for the discrete transition is weaker than the continuous one. On the other hand, the interference features dissolve in the case that the transition to the discrete level is enlarged with respect to the continuous transition, and the trace of the interference derived from four-wave mixing recovers the shape of those produced due to a superposition of a damped resonance and a continuous one.

In the case of discrete states coupled to a continuous state, the pulses can be established through an underdamped state, which is determined through the separation of the broadened spectral peaks. Also, differences are established in the linear polarization spectrum of the overdamped case due to the analogous time of the narrowing of the Fano resonances. Therefore, it is interesting to analyze interference phenomena through the four-wave mixing technique through atomic or molecular systems, as well as semiconductor heterostructures through which the design of excitonic levels is possible [22].

The four-wave mixing is closely related to the third-order susceptibility response, χ^3 such that, to establish Fano resonance through this technique, two plasmon modes with frequencies ω_1 and ω_2 are pumped and as a result a different frequency emerges in another mode. Hence, two plasmons ω_1 are eliminated and two plasmons of different frequencies ω_2 and ω_3 are created. The above described can be defined as

$$\widehat{H}_{\text{FMW}} = \left(\int d^3(r)E_3^*(r)\chi_3(r)E_2^*(r)E_1^2(r) \right)\widehat{a}_3^\dagger\widehat{a}_2^\dagger\widehat{a}_1^\dagger + H.c. \qquad (2.23)$$

Through this technique, it is possible to explore third-order optical nonlinearities, such that the design of nonlinear Fano resonances is possible [17].

2.4.4 Influence of Fano Resonances in Raman Scattering

Raman spectroscopy is one of the most widely used measurement techniques for the characterization of nanostructures since it is one of the methods for the identification of the number of layers of multilayer materials. Moreover, Raman spectroscopy is very useful in the study of specific characteristics such as doping, deformation, and mechanical properties of nanostructures.

Localized plasmon polarization interacts with the vibrational modes of a molecule. In the Stokes shift, a plasmon with frequency ω is absorbed by the molecule, which also generates a vibrational mode and a lower-energy plasmon. So, through the frequency generated by the plasmon, a significant Raman intensity is produced. This process is described with the following equation: the Hamiltonian for the process can be written similarly to

$$\widehat{H}_{\text{R}} = \left(\int d^3(r)\rho(r)E_1^*(r)E_2(r) \right)\widehat{b}^\dagger\widehat{a}_1^\dagger\widehat{a}_2 + H.c. \qquad (2.24)$$

where $b^{\hat{\dagger}}$ creates a phonon in the vibratory mode of the molecule. The density of the molecule $\rho(r)$ can be considered as a 3D step function. On the other hand, the Hamiltonian can result from an optomechanical coupling between the oscillations of the molecule and the plasmon.

Fano resonances have been observed in Raman spectra in carbon nanostructures. Fano resonances have been observed in infrared spectra in the presence of a bandgap of energy generated by a perpendicular electric field. This is due to an inference between the phonon and a continuous spectrum of electronic transitions near the bandgap of the material. Further, in Fano resonances in low-frequency cutoff phonon modes, Raman spectra have also been observed and have been interpreted as the interference between C-mode phonons and continuous electronic transitions.

Furthermore, there is a dependency between the Fano resonances in the infrared spectra due to the layer thickness of the multilayer nanomaterial. While, for single-layer nanostructures, Fano resonances have not been observed in Raman scattering.

2.5   Theoretical Models for Describing the Influence of Bound States in Optical Nonlinearities

Optical bound states have attracted attention in recent times since through them it is possible to generate resonant responses with extremely narrow spectral characteristics. Equally, the bound states are peculiar because their treatment depends on the classical or quantum regimen.

The above is demonstrated because the probability of a particle to cross a potential barrier in a quantum problem is not zero. However, in a classical problem, it is possible to consider bound and unbound states. Besides, a bound state in the quantum regime is directly related to the notion of time because the particles that make up the system must always remain localized. The foregoing corresponds to configurations that remain localized for an infinite time and maintain constant energy, such that this state is decomposed as a combination of bound states. What corresponds to a localized state is a set of linked states. In addition, there are states that remain confined for a finite time after delocalizing. On the other hand, the localization time refers to the time it takes for the particle to cross the potential barrier that kept it confined.

On the other hand, because bound states are decoupled through symmetry protection or resonance trapping, they can be resistant to perturbations of the geometric parameters of the photonic crystal. Also, because bound states provide an efficient light capture mechanism, optical bound states are particularly attractive for enhancing nonlinear optical effects.

Also, one of the ways to improve the response of optical nonlinearities is through excitons–polaritons and hybrid quasiparticles, which obtain the coherent properties of photonic modes such as the interaction force of excitons. Thus, hybrid nanophotonic systems incorporating transition metal dichalcogenides are promising structures due to their ability to operate at room temperature. Moreover, the coupling of transition metal dichalcogenides excitons to optical bound states in photonic crystals favors an increase in optical nonlinearities, in addition to providing control over the properties of resonant bound states through the excitonic fraction in the polariton [23]. Furthermore, it has been shown that bonded optical states can be generated through an optical waveguide array by decoupling relative to waveguide symmetry [24].

The energies and wave functions of the bound states are obtained by the Schrödinger equation

$$\widehat{H}\psi(r,z,\phi) = E\psi(r,z,\phi) \tag{2.25}$$

So, for bound-state solutions, the relevant factor for the optical response is expressed through [25]

$$|\psi_n(r=0)|^2 = \frac{\lambda^3}{32\pi^2}g^3\frac{1}{n}\left(\frac{1}{n^2}-\frac{n^2}{g^2}\right) \tag{2.26}$$

On the other hand, bound states based on Mie resonances have been used in order to improve nonlinear optical processes. It has been shown that through crystalline nanostructures, and using bound states, improvements in second-order optical nonlinearities of up to three orders of magnitude have been achieved since through bound states symmetric defects are introduced that allow transitions between bound states. Equally, the polarization characteristics of the second harmonic generation establish an improvement that is attributed to the quantum bound-state mode. Thus, before breaking the structural symmetry of the nanoresonators arranged in lattice structures, high Q factor resonances can be obtained from the bound states.

2.6   Theoretical Models for Describing the Influence of the Purcell Effect in Optical Nonlinearities

The Purcell effect is derived from a system coupled to an electromagnetic resonator, the probability of spontaneous emission increases over its global value, and the recombination time is reduced by a factor described as [2]

$$F_{\mathrm{P}} = \frac{3\lambda_{\mathrm{C}}^3}{4\pi^2}\frac{Q}{V} \tag{2.27}$$

where V is related to the volume of the resonant mode, Q refers to the quality factor, and λ_{c} is the wavelength in the material that is also related to the refractive index of the medium. So, for the increase in the emission rate, the confinement of light through optical resonators to dimensions comparable to the wavelength is necessary; in addition, a considerable storage time is necessary.

On the other hand, the Purcell effect determines the understanding of nanolaser models. The Purcell factor, and the spontaneous emission coupling factor referred to as β, is set as adjustable parameter. The β and F_{p} parameters are also referred to as the result of the emission processes that occur in a nanolaser. Since the active medium of nanolasers consists of semiconductors and multiquantum wells, homogeneous and inhomogeneous broadenings must be considered, in the case of nanolasers as a function of room temperature.

The study of the Purcell effect has stood out because it offers a field of study based on the study of the optical response to complex dielectric environments. It can improve or suppress the spontaneous emission of a dipole source. Thus, the Purcell effect can also be used for the adaptation of nonlinear optical susceptibilities [26]. Therefore, through the Purcell effect, it has been shown that it is possible for frequencies close to an atomic resonance to influence the resulting nonlinearity of Kerr for light of all frequencies. One way of explaining the modification of the optical nonlinearities through the Purcell effect is through the optical nonlinearities derived from atomic resonances, so that the modification of their strengths in turn influences the strengths of the nonlinearities.

On the other hand, the influence of the Purcell effect on the response in optical nonlinearities can be started through a two-level optical system such as the Kerr effect. Derived from the above, we can start by considering a series of N systems of two levels per unit volume in a cavity based on photonic crystals, whose levels are labeled a and b. So, the Hamiltonian corresponds to the sum of the interaction and energy terms. Therefore, the approximation of the electric dipole can be expressed as

$$H = \hbar[\omega_a\sigma_{aa} + \omega_b\sigma_{bb} + \Omega(t)\sigma_{ab} + \Omega^*(t)\sigma_{ba}] \tag{2.28}$$

where σ_{ij} is the operator that transforms the fermionic state j to the fermionic state i, $\Omega(t)$ is the amplitude Rabi of the applied field as a function of time, and μ refers to the scalar dipole moment that is defined in terms of its projection along the length of the applied field $E(t)$.

The system described above is also defined by the environmental degrees of freedom, such that there is a loss of information sent to the environment; this is also known as the Markov approach. This system can also be explained in terms of the following equation:

$$\dot{\rho} = -\left(\frac{i}{\hbar}\right)[H,\rho] + \sum\left[L_\mu\rho L_\mu^\dagger - \frac{1}{2}L_\mu^\dagger L_\mu\rho - \frac{1}{2}L_\mu^\dagger L_\mu\right] \tag{2.29}$$

If two quantum jump operators are taken into account, it is possible to obtain the following dynamic equations:

$$\frac{d\rho_{ba}}{dt} = -\left(i\omega_{ba} + T_2^{-1}\right)\rho_{ba} + i\Omega(t)(\rho_{bb} - \rho_{aa}) \tag{2.30}$$

$$\frac{d(\rho_{bb} - \rho_{aa})}{dt} = \frac{(\rho_{bb} - \rho_{aa}) + 1}{T_1} - 2i[\Omega(t)\rho_{ab} - \Omega*(t)\rho_{ba}] \tag{2.31}$$

Then, to define the influence of the Purcell effect on optical nonlinearities near resonance, a rotating wave approximation can be used from previous equations where, if the polarization is defined by $P = N\mu(\rho_{ba}-\rho_{ab}) = \chi^E$, and also, χ refers to the total susceptibility to all orders; it is possible to obtain the following equation to define the optical susceptibility:

$$\chi = -\frac{N\mu^2\left(\omega - \omega_{ba} - \frac{1}{T_2}\right)T_2^2/\hbar}{1 + (\omega - \omega_{ba})^2 T_2^2 + (4/\hbar^2)\mu^2|E|^2 T_1 T_2} \tag{2.32}$$

By expanding in powers of the electric field squared the above equation, it can be determined that

$$\chi^{(3)} = \frac{4}{3}N\mu^4 \frac{T_1 T_2^2(\Delta T_2 - i)}{\hbar^3\left(1 + \Delta^2 T_2^2\right)^2} \tag{2.33}$$

If $\Delta \equiv \omega-\omega_{ba}$ is considered to be the detuning of the incoming wave of the electron resonance frequency, then, for large detunings, the following approximation is obtained:

$$\operatorname{Re}\chi^{(3)} \approx \frac{4}{3}N\mu^4 \left(\frac{1}{\hbar\Delta}\right)^3 \frac{T_1}{T_2} \tag{2.34}$$

On the other hand, if the real coefficient of third-order optical susceptibility is defined in terms of the energy of the bandgap of the system, then the following equation can be obtained:

$$\operatorname{Re}\chi^{(3)} \approx \frac{1}{30\pi\hbar^3}\left(\frac{eP}{\hbar\omega}\right)^4 \left|\frac{2m_r}{\hbar\Delta}\right|^{3/4} \frac{T_1}{T_2} \tag{2.35}$$

where P is an element of the matrix, ωG is the direct bandgap energy of the system, and mr is the reduced effective mass of the exciton. So, finally, the improvement of $\operatorname{Re}\chi^{(3)}$ derived from the Purcell effect is defined through the following equation:

$$\eta \approx \frac{T_{1,\text{Purcell}}\, T_{2,\text{vac}}}{T_{1,\text{hom}}\, T_{2,\text{Purcell}}} = \frac{\frac{1}{2}\Gamma_{nr} + \Upsilon_{\text{phase}}}{\frac{1}{2}(\Gamma_{nr} + \Gamma_{\text{rad}}) + \Upsilon_{\text{phase}}} \frac{\Gamma_{\text{rad}} + \Gamma_{nr}}{\Gamma_{nr}} \tag{2.36}$$

where Γ_{rad} is the radiative decay rate in vacuum.

2.7 Theoretical Models for Describing the Contribution of Different Physical Mechanisms Responsible for an Enhancement in Nanoscale Nonlinear Optical Effects

There are different physical mechanisms responsible for modifying the nonlinear optical response of nanostructures. The physical mechanisms mentioned above refer to electronic polarization, molecular orientation, electrostriction, saturable absorption, thermal effects, redistribution of free electrons, and photorefractive effect. However, there are more physical mechanisms that produce a negligible contribution to the optical response exhibited by nanomaterials.

2.7.1 Nonlinear Polarization of the Medium by the Bonding Electrons

This physical mechanism occurs through a distortion in the cloud of bonding electrons in the atoms or molecules in the environment that occurs due to the intensity of the incident electric field. Furthermore, this mechanism also causes an electrical polarization as a nonlinear response to the stimulus of the electromagnetic field. Also, nonlinearity depends on the facility for an atom or molecule to be polarized. This mechanism is so fast that it is considered almost instantaneous. So, the response time is of utmost importance for the distortion of the electronic cloud in the presence of an optical field. So, the response time is estimated through the orbital period of an electron as it travels around the nucleus, which can be expressed as

$$\tau = \frac{2\pi a0}{ve} \tag{2.37}$$

where $a_0 = 0.5 \times 10^{-10}$ m refers to the Bohr radius of the atom, and $v_e \approx c/137$ is a typical electronic velocity, with which $\tau \approx 0.1$ fs.

2.7.2 Generation of Free Carriers

This mechanism consists of the energy gain of an electron to be transferred from the valence band to the conduction band for every two photons absorbed through a semiconductor. Thus, the modification in the density of free carriers produces a change in the refractive index of a system. Also, the electrons in the conduction band return to their basal state by recombining with the holes left in the valence band such that the contribution of this physical mechanism to the nonlinear refractive index n_2 is given by

$$n_2 = K \frac{hc\sqrt{E_P}}{2n_0^2 E_g^4} G_2\left(\frac{\hbar\omega}{E_g}\right) \tag{2.38}$$

where $E_p = 21 \ eV$ defined for direct semiconductors, $K = 3.1 \times 10^3$ eV associated with an independent parameter of the material, and G_2 is the universal function that is expressed as

$$G_2(x) = \frac{-2 + 6x - 3x^2 - x^3 - \frac{3}{4}x^4 - \frac{3}{4}x^5 + 2(1 - 2x)^{\frac{3}{2}} \Theta(1 - 2x)}{64x^6} \tag{2.39}$$

where Θ is the Heaviside step function, and, on the other hand, the nonlinear coefficient related to two-photon absorption is given by

$$\beta = K \frac{\sqrt{E_P}}{n_0^2 E_g^3} F_2\left(\frac{2\hbar\omega}{E_g}\right) \tag{2.40}$$

where F_2 is the universal function defined through

$$F_2(2x) = \frac{(2x - 1)^{\frac{3}{2}}}{(2x)^5} \tag{2.41}$$

On the other hand, the absorption of two photons disappears for $\hbar_\omega < Eg/2$ due to the bandgap energy and the energy of the incident photon. Like the linear refractive index n_0, the nonlinear refractive index n_2 presents dispersion, and for the case of a direct semiconductor, the value of n_2 is positive for long wavelengths, with a reinforcement around the band of absorption being nonlinear and becomes negative for short wavelengths.

2.7.3   Molecular Orientation

Molecular orientation improves the physical and mechanical properties of nanostructures without altering the benefits and chemical properties of the original polymer. The origin of the optical nonlinearities is related to the alignment of the molecules in the electric field of an applied optical wave such that the optical wave then undergoes a modified value of the refractive index due to the modification of the mean polarization per molecule through the molecular alignment.

For this mechanism, it is necessary to consider the nonlinear response of anisotropic molecules to light of arbitrary polarization. Thus, through this mechanism, the modification of the second-order nonlinear refractive index is observed through

$$\bar{n}_2 = \frac{N}{45n_0}\left(\frac{n_0^2+2}{3}\right)^4\frac{(\alpha3-\alpha1)}{kT} \qquad (2.42)$$

where α_1 and α_3 refer to polarizability, N is the number density of molecules, k refers to the Boltzmann's constant, and T is the thermodynamic temperature.

2.7.4 Electrostriction

Electrostriction is the tendency of materials to compress in the presence of an electric field. This mechanism encourages nonlinear third-order optical response. So, the electrostriction force is related to a consequence of maximizing stored energy. Also, the potential energy changes from its value without the field by the amount, as seen:

$$u = \frac{1}{2}\epsilon\epsilon_0 E^2 \qquad (2.43)$$

where ϵ is the relative dielectric constant of the material, and ϵ_0 refers to the allowance of free space.

On the other hand, the modification of the optical properties is the result of electrostriction. So, the change in susceptibility in the presence of an optical field is represented as $\Delta\chi = \Delta\epsilon$, where $\Delta\epsilon$ is calculated as $(\partial\epsilon/\partial\rho)\,\Delta\rho$, such that where ϵ is the relative dielectric constant of the material and ϵ_0 refers to the allowance of free space.

$$\Delta\chi = \epsilon_0 C_T\gamma_e^2 \boldsymbol{E}\cdot\boldsymbol{E}^* \qquad (2.44)$$

Thus, due to electrostriction, it is possible to establish its influence on the optical nonlinearities of a system as

$$\chi^{(3)}(\omega=\omega+\omega-\omega) = \frac{\epsilon_0\gamma}{3\nu^2}\left[\frac{(n^2-1)(n^2+2)}{3}\right]^2 \qquad (2.45)$$

2.7.5 Saturable Absorption

An example of a nonparametric nonlinear optical process is saturable absorption. The said mechanism consists of the behavior of the absorption coefficient of a system, which decreases when measurements are made with a high intensity such that the dependence of the measured absorption coefficient α on the intensity I of the incident laser radiation is expressed through

$$\alpha = \frac{\alpha_0}{1 + I/I_S} \tag{2.46}$$

where α_0 is the low-intensity absorption coefficient, and $I_{s,}$ refers to the saturation intensity.

2.7.6 Thermal Effects

The physical mechanisms associated with thermal processes generally exhibit large nonlinear optical effects. Thus, nonlinear thermal optical effects originate through the incident laser in the nanostructured system because a fraction of the incident laser power is absorbed as it passes over the material. Consequently, the illuminated part of the material exhibits a significant increase in temperature, which leads to a change in the nonlinear refractive index of the material. Moreover, the thermal effects that cause nonlinear optical phenomena are highly dependent on time. So, it can be assumed that the refractive index $\tilde{n}$ varies through temperature and can be expressed as

$$\tilde{n} = n_0 + \left(\frac{dn}{dT}\right)\tilde{T}_1 \tag{2.47}$$

where (dn/dT) defines the temperature dependence of the refractive index, and T_1 designates the temperature change induced by a laser pulse.

On the other hand, the influence of thermal effects on the nonlinear optical response is different for a continuous light beam and for a pulsed light beam. So, for the response between a continuous wave light beam, it is considered that under steady-state conditions, the heat transport equation is expressed as

$$-\kappa \nabla^2 \tilde{T}_1 = \propto \widetilde{I(r)} \tag{2.48}$$

Through the above, it is possible to estimate the nonlinear optical change through the refractive index expressed as

$$n_2^{(\text{th})} = \left(\frac{dn}{dT}\right)\frac{\propto R^2}{\kappa} \tag{2.49}$$

On the other hand, it is necessary to define that the change in the refractive index also depends on the geometry so that it is not considered as an intrinsic property of an optical system.

On the other hand, for pulsed lasers, the induced change in the re-fraction index is proportional to the pulse energy $Q = \int P(t)\, dt$. Thus, through this mechanism, the refractive index increases or decreases monotonically during the time span of the laser pulse. So, the condition under which there is a change in the refractive index due to this thermal mechanism is given by

$$\Delta n^{(\text{th})} = \left(\frac{dn}{dT}\right) T_1^{(\max)} \tag{2.50}$$

2.8 Highlights of Theoretical Models for an Enhancement in Optical Nonlinearities

It has been defined that the field of study of the nonlinear regime due to the interaction between light and matter expanded due to the development of lasers. Nonlinear optics has been shown to enable numerous techniques for characterizing nanostructures, such as harmonic generation. Because there is a weakness in the nonlinear optical response of conventional materials, nanostructures show considerable nonlinear optical effects at reasonable light intensities. Therefore, a long propagation length in a material is necessary to achieve significant nonlinear effects.

On the other hand, thermal mechanisms also contribute to the nonlinear response of nanostructures. In addition, the generation of second- and third-order harmonics, in addition to the multiwave mixing technique, is a suitable experimental measurement for order-of-magnitude improvement of optical nonlinearities. Moreover, further improvement in optical nonlinearities can be obtained through path interference effects due to Fano resonances. Furthermore, quantum confinement and optical nonlinearities in combination are of importance for applications associated with light modulation. Besides, optical modulation and switching of light are also possible through nonlinearity induced by field location. In addition, the presence of one of the pulses can cause changes in the refractive index of the medium by inducing a Kerr-type nonlinearity.

Moreover, bound states have been used to improve lower-order harmonics, as well as higher optical harmonic generations for photonic device design. Thus, through the application of bound states, the light–matter interaction has potential for the design of techniques for solid-state attosecond spectroscopy and UV sources. Thus, through bound states, advances in nonlinear optics could be generated in addition to a wide variety of application fields, such as nonlinear microscopy, optical computation, and signal processing, as well as quantum photonic technologies. Lately, quantum confinement effects have been examined in order to exploit optical properties for the design of applications in the field of quantum information processing, mainly quantum bits and nonlinear quantum operations.

With this it is possible to establish the study area focused on integrated quantum photonics; its main problem is the photon interaction force, resulting in optical nonlinearities too small for applications intended for nonlinear optical quantum, even in the best nonlinear crystals. The polaritons can be essentially described as clothed photons, inheriting an effective light mass along with a strongly interacting nucleus. Thus, excitons could be established as ideal for powering nonlinear quantum gates for the development of photonic quantum devices.

References

1. Chemla, D. S. (1993). Nonlinear optics in quantum-confined structures. *Physics Today, 46*(6), 46–52.
2. Flytzanis, C., Hache, F., Klein, M. C., & Ricard, D. (1990). Impact of quantum confinement on the optical nonlinearities of semiconductor nanocrystals. *Optics in Complex Systems, 1319,* 75–80. International Society for Optics and Photonics.
3. Banyai, L., Hu, Y. Z., Lindberg, M., & Koch, S. W. (1988). Third-order optical nonlinearities in semiconductor microstructures. *Physical Review B, 38*(12), 8142.
4. Roussignol, P. H., Ricard, D., & Flytzanis, C. H. R. (1990). Quantum confinement mediated enhancement of optical kerr effect in CdS x Se 1− x semiconductor microcrystallites. *Applied Physics B, 51*(6), 437–442.
5. Gerace, D., Laussy, F., & Sanvitto, D. (2019). Quantum nonlinearities at the single-particle level. *Nature Materials, 18*(3), 200–201.
6. Taghizadeh, A., Thygesen, K. S., & Pedersen, T. G. (2021). Two-dimensional materials with giant optical nonlinearities near the theoretical upper limit. *ACS Nano, 15*(4), 7155–7167.
7. Priyadarshini, M., Acharyya, J. N., Mahajan, S., & Prakash, G. V. (2021). Optical nonlinearities in chemically synthesized and femtosecond laser fabricated gold nanoparticle colloidal solutions. *Optics & Laser Technology, 139,* 107008.
8. Ricard, D., Roussignol, P., Hache, F., & Flytzanis, C. (1990). Nonlinear optical properties of quantum confined semiconductor microcrystallites. *Physica Status Solidi (b), 159*(1), 275–285.
9. Keilmann, F., & Hillenbrand, R. (2004). Near-field microscopy by elastic light scattering from a tip. *Philosophical Transactions of the Royal Society of London. Series A: Mathematical, Physical and Engineering Sciences, 362*(1817), 787–805.
10. Kauranen, M., & Zayats, A. V. (2012). Nonlinear plasmonics. *Nature Photonics, 6*(11), 737–748.
11. Vasilantonakis, N., Wurtz, G. A., Podolskiy, V. A., & Zayats, A. V. (2015). Refractive index sensing with hyperbolic metamaterials: Strategies for biosensing and nonlinearity enhancement. *Optics Express, 23*(11), 14329–14343.
12. Boyd, R. W., Shi, Z., & De Leon, I. (2014). The third-order nonlinear optical susceptibility of gold. *Optics Communications, 326,* 74–79.
13. Piredda, G., Smith, D. D., Wendling, B., & Boyd, R. W. (2008). Nonlinear optical properties of a gold-silica composite with high gold fill fraction and the sign change of its nonlinear absorption coefficient. *JOSA B, 25*(6), 945–950.
14. Gaponenko, S. V. (2010). *Introduction to nanophotonics.* Cambridge University Press.
15. Reyna, A. S., & de Araújo, C. B. (2017). High-order optical nonlinearities in plasmonic nanocomposites-a review. *Advances in Optics and Photonics, 9*(4), 720–774.
16. Rau, A. R. P. (2004). Perspectives on the Fano resonance formula. *Physica Scripta, 69*(1), C10.
17. Thyagarajan, K., Butet, J., & Martin, O. J. (2013). Augmenting second harmonic generation using Fano resonances in plasmonic systems. *Nano Letters, 13*(4), 1847–1851.

18. Limonov, M. F., Rybin, M. V., Poddubny, A. N., & Kivshar, Y. S. (2017). Fano resonances in photonics. *Nature Photonics, 11*(9), 543–554.
19. Luk'yanchuk, B., Zheludev, N. I., Maier, S. A., Halas, N. J., Nordlander, P., Giessen, H., & Chong, C. T. (2010). The Fano resonance in plasmonic nanostructures and metamaterials. *Nature Materials, 9*(9), 707–715.
20. Turkpence, D., Akguc, G. B., Bek, A., & Tasgin, M. E. (2014). Engineering nonlinear response of nanomaterials using Fano resonances. *Journal of Optics, 16*(10), 105009.
21. Butet, J., & Martin, O. J. (2014). Fano resonances in the nonlinear optical response of coupled plasmonic nanostructures. *Optics Express, 22*(24), 29693–29707.
22. Meier, T., Schulze, A., Thomas, P., Vaupel, H., & Maschke, K. (1995). Signatures of Fano resonances in four-wave-mixing experiments. *Physical Review B, 51*(20), 13977.
23. Kravtsov, V., Khestanova, E., Benimetskiy, F. A., Ivanova, T., Samusev, A. K., Sinev, I. S., & Iorsh, I. V. (2020). Nonlinear polaritons in a monolayer semiconductor coupled to optical bound states in the continuum. *Light: Science & Applications, 9*(1), 1–8.
24. Molina, M. I., Miroshnichenko, A. E., & Kivshar, Y. S. (2012). Surface bound states in the continuum. *Physical Review Letters, 108*(7), 070401.
25. Haug, H., & Koch, S. W. (2009). *Quantum theory of the optical and electronic properties of semiconductors*. World Scientific Publishing Company.
26. Iwase, H., Englund, D., & Vučković, J. (2010). Analysis of the Purcell effect in photonic and plasmonic crystals with losses. *Optics Express, 18*(16), 16546–16560.

Chapter 3
Study on Second- and Third-Order Nonlinear Optical Properties in Nanostructured Carbon Allotropes

3.1 Carbon Allotropes: Structure, Properties, and Applications

Carbon-focused nanoscience has attracted special interest in recent years. Research in this field involves studying the properties of groups of carbon entities at the nanoscale. In the same way, importance has been given to the study of the conditions of geometric transformation of carbon nanostructures and the possibility of synthesizing hybrid carbon nanomaterials. In addition, an extensive field of nanotechnology perspectives focuses on carbon-based nanostructures.

Carbon nanomaterials have a wide range of geometric structures, such as diamonds, graphite, graphene, and nanotubes. Carbon is an element with a polyform characteristic, that is, it has the ability to exist in more than one crystalline structure. Moreover, the polymorphic classification of carbon is called allotropy since it is made up of a single element. Each polyform has physical and chemical properties that vary remarkably according to its structure, such as density, hardness, and shape.

Characteristic geometries assign very good optoelectronic properties to carbon nanostructures. The two-dimensional carbon materials also stand out since they can be described through the Dirac equation in addition to having an electronic structure that highlights the interactions of electromagnetic waves with matter. In addition, it has been highlighted that the structures of carbon nanomaterials are related to each other, which allows the transformation between one geometry and another.

On the other hand, in the field of nonlinear optical study of carbon samples, the use of exploration techniques through coupling of optical waves is convenient. In the case of carbon nanostructures, the analysis of the optical absorption spectra is necessary; in addition, it is important to consider excitonic effects at the nanoscale. Besides, it is possible to pose the effects of electrons in their absorption spectra as shown in Fig. 3.1.

© The Author(s), under exclusive license to Springer Nature Switzerland AG 2022

C. Torres-Torres, G. García-Beltrán, *Optical Nonlinearities in Nanostructured Systems*, Springer Tracts in Modern Physics 287, https://doi.org/10.1007/978-3-031-10824-2_3

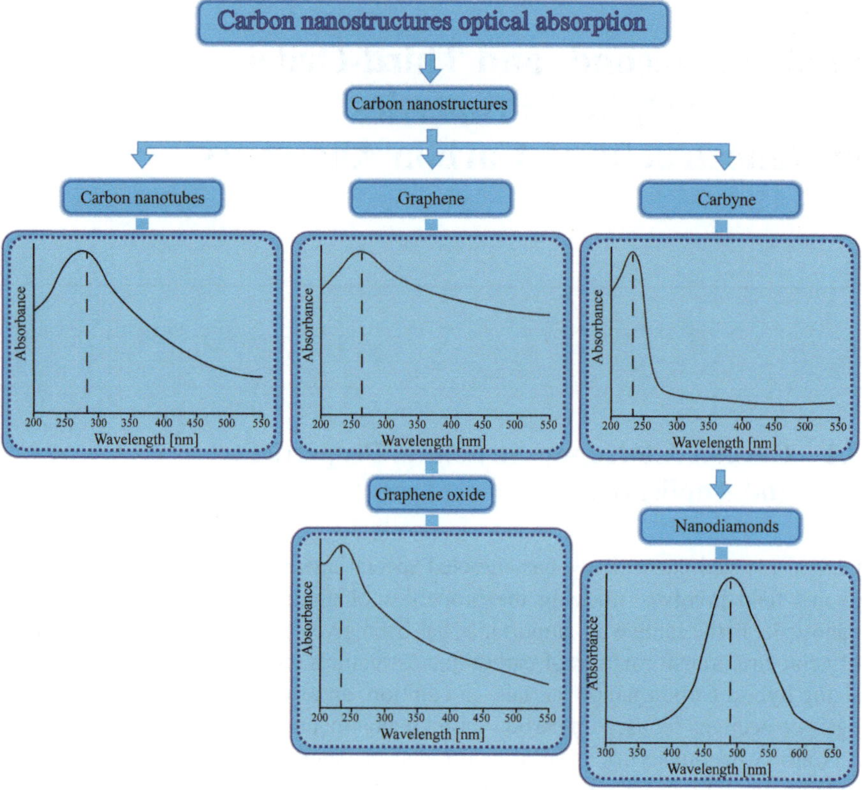

Fig. 3.1 Characteristic absorption spectra of metallic nanostructures

Consequently, the structures and properties of carbon nanostructures make them potential materials for technological applications in different fields of research. Thus, carbon-based nanostructures are excellent candidates for sensors and optoelectronic components in nanoscale circuits. On the other hand, the field of carbon nanoscience continues to expand with the discovery of new carbon allotropes.

3.1.1 Graphite

Graphite is a polymorphic state of carbon made up of overlapping sheets of hexagonal structure, as shown in Fig. 3.2.

The hexagonal structure is formed due to the $C = C$ double bonds in the same plane at an angle of $120°$. The covalent bonds between single-layer carbon atoms are very strong. However, the bonds between the layers are weak due to van der Waals forces.

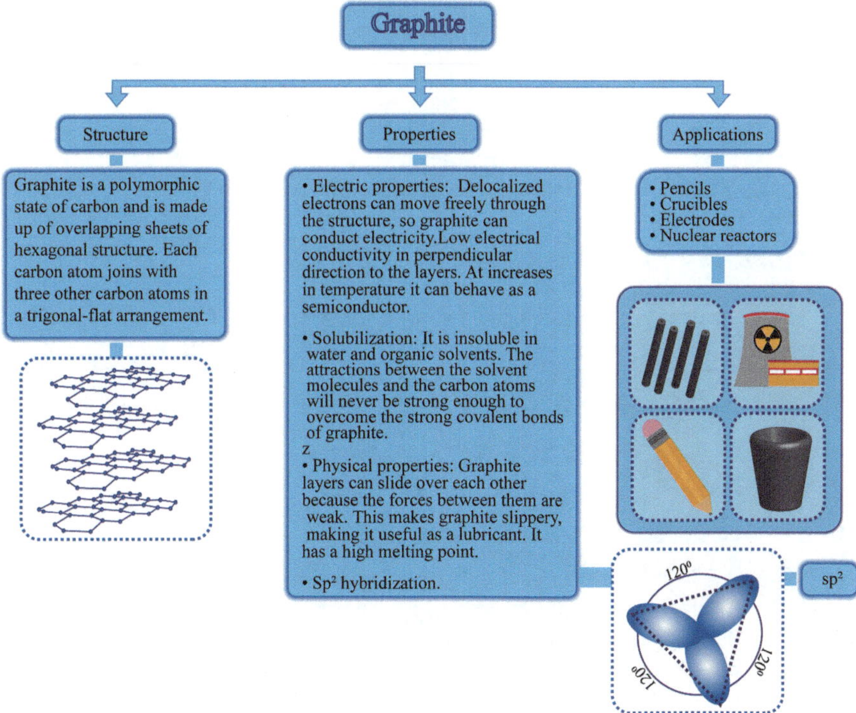

Fig. 3.2 Graphical synthesis of graphite properties and applications

Furthermore, graphite is the most stable allotropic form of carbon. Graphite is black, shiny, and opaque, not transparent, and also a very slippery material. However, it is very difficult to melt and takes a great deal of energy to break the strong covalent bonds and separate the carbon atoms from the graphite.

3.1.2   Fullerenes

Fullerenes are pure carbon macromolecules that can vary in shape depending on the number of carbon atoms that compose them, as is described in Fig. 3.3.

Such shapes can be spherical, ellipsoidal, or tubular. Its structure is due to each carbon atom being linked to three others. Carbon atoms are linked together by three very strong covalent bonds, one double and two single bonds. The structure of the fullerenes is also highly symmetrical and stable.

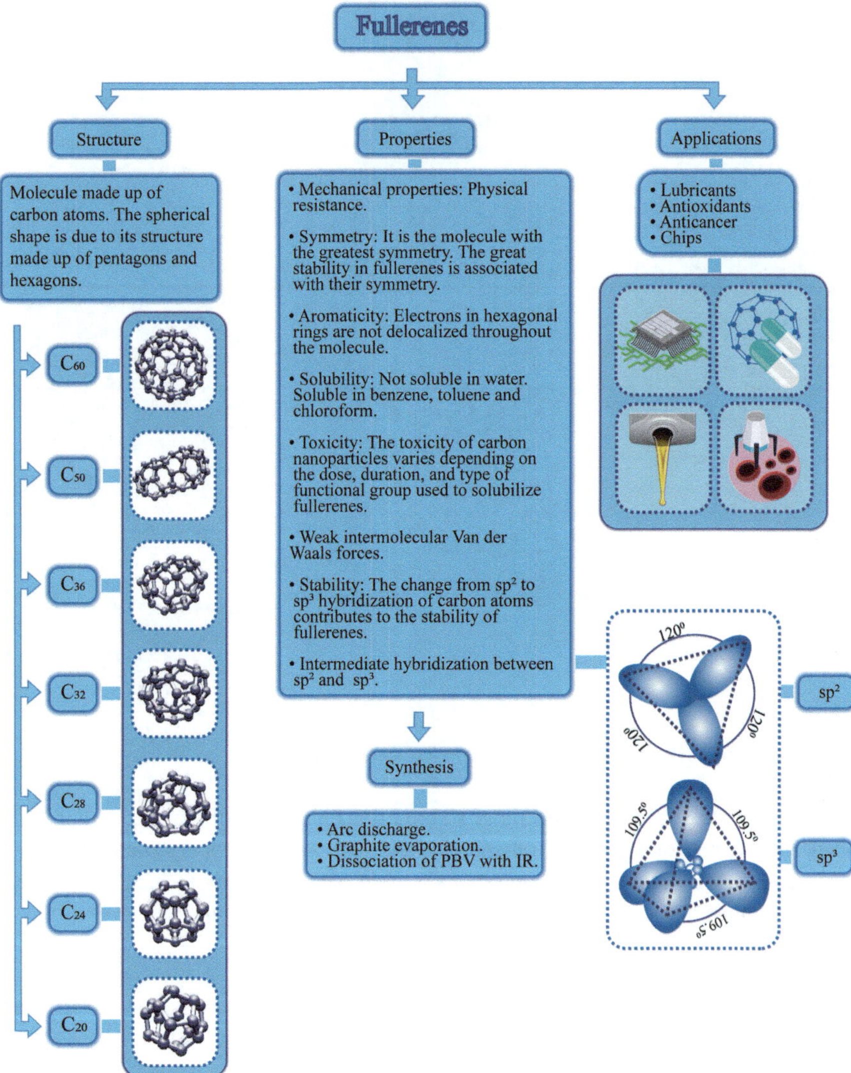

Fig. 3.3 Graphical synthesis of fullerenes properties and applications

3.1.3   Carbon Nanotubes

The structure of carbon nanotubes consists of carbon molecules connected to three neighboring atoms through covalent bonds, as described in Fig. 3.4.

The C = C double bonds formed in carbon nanotubes provide π-electrons around the walls of carbon tubes. Thus, the layers of the structure of multiwall carbon nanotubes allow such aspects to be considered in nonlinear optical studies.

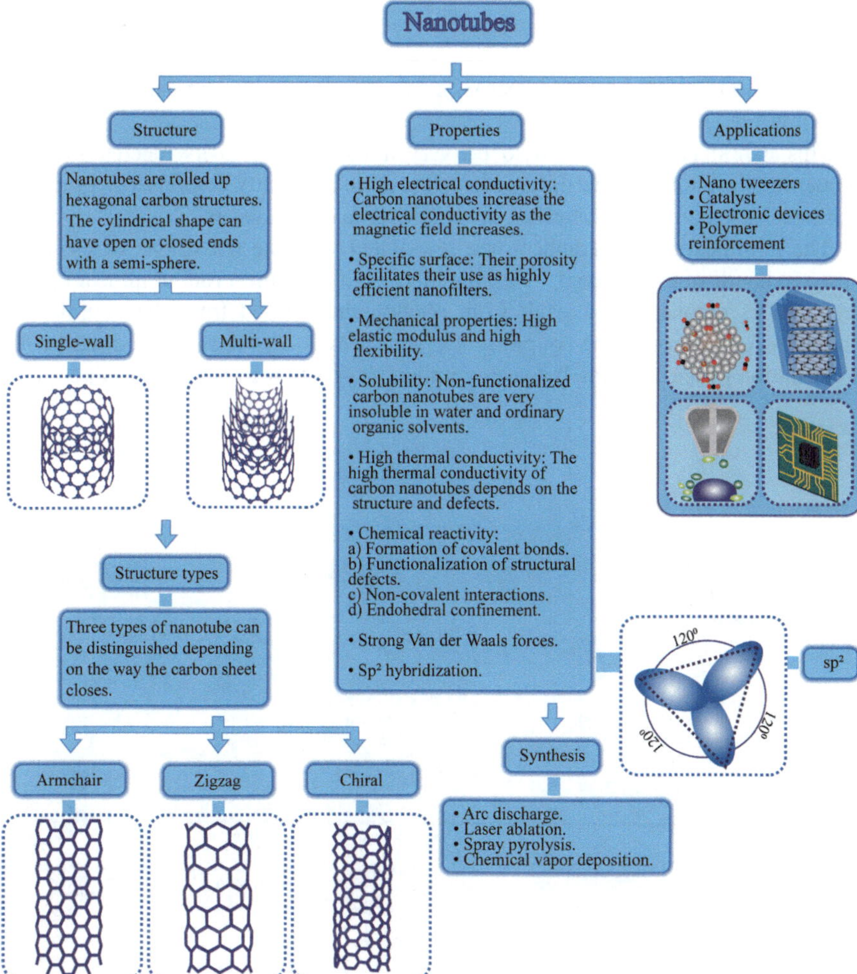

Fig. 3.4 Graphical synthesis of carbon nanotubes properties and applications

Carbon nanotubes have specific characteristics due to their structure. Since each carbon atom is bonded to three other carbon atoms through strong covalent bonds, it has a high melting point. This also forms a high number of delocalized electrons within the nanotube, so each carbon atom has a spare electron, resulting in the excellent electrical conductivity of carbon nanotubes.

3.1.4   Graphene

Graphene consists of a carbon sheet with two-dimensional hexagonal geometry, as shown in Fig. 3.5.

This nanomaterial turns out to be very attractive due to its mechano-optical properties. In addition, the third-order optical susceptibility response of graphene exhibits outstanding nonlinearities in multilayer and monolayer.

The nonlinearities of graphene are associated with the transitions of electrons between bands of the material. Thus, the increase in the third-order nonlinear optical response of graphene can result from phenomena derived from the coupling of optical waves. Through the coupling of optical waves, it is possible to describe the contribution of the coupling between graphene layers to the resulting nonlinearities.

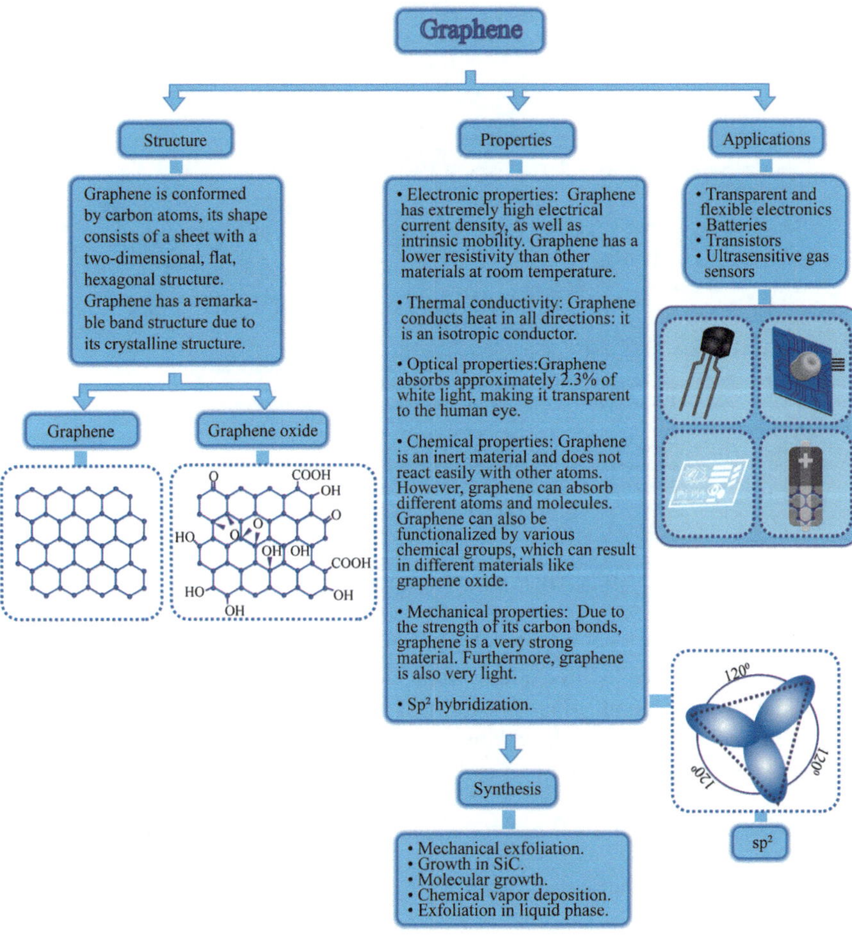

Fig. 3.5 Graphical synthesis of graphene properties and applications

3.1.5 *Graphane*

On the other hand, graphene has a monolayer structure similar to graphene, with the difference that the carbon atoms, in addition to being linked to each other, are also linked to hydrogen atoms located on both sides of the layer as observed in Fig. 3.6.

3.1.6 *Carbyne*

Carbyne, described in Fig. 3.7, is a chain of carbon atoms linked either by alternating triple and single bonds or by consecutive double bonds. Its properties are because carbon unfolds differently. In this case, the carbon atoms are presented in the form of a one-dimensional chain and form a network very similar to that of diamond.

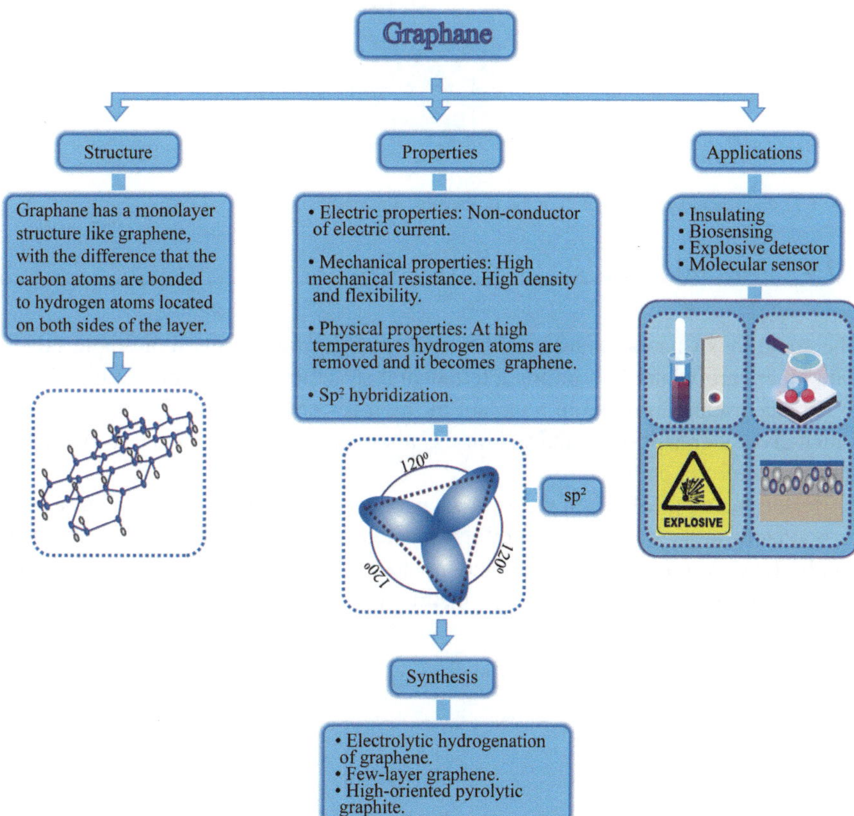

Fig. 3.6 Graphical synthesis of graphane properties and applications

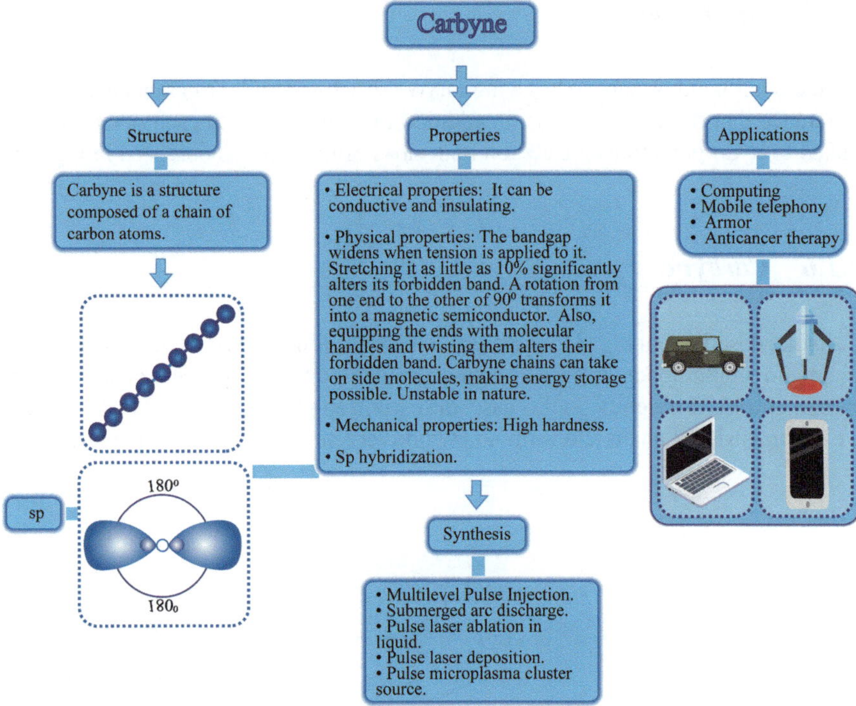

Fig. 3.7 Graphical synthesis of carbyne properties and applications

3.1.7 Nanodiamonds

The structure of nanodiamonds is that each carbon atom forms four single bonds with other carbon atoms, forming a tetrahedral structure, as described in Fig. 3.8.

Tetrahedra are arranged in a variant of the face-centered cubic crystal structure, which gives it greater order and stability. Carbon nanodiamonds are the second most stable structure in carbon.

Besides being insoluble in water, it is a poor conductor of electricity because each carbon atom in the crystal structure of a diamond is linked by four strong covalent bonds, leaving no free electrons and no ions. This property also indicates the hardness of the diamond and its high melting point.

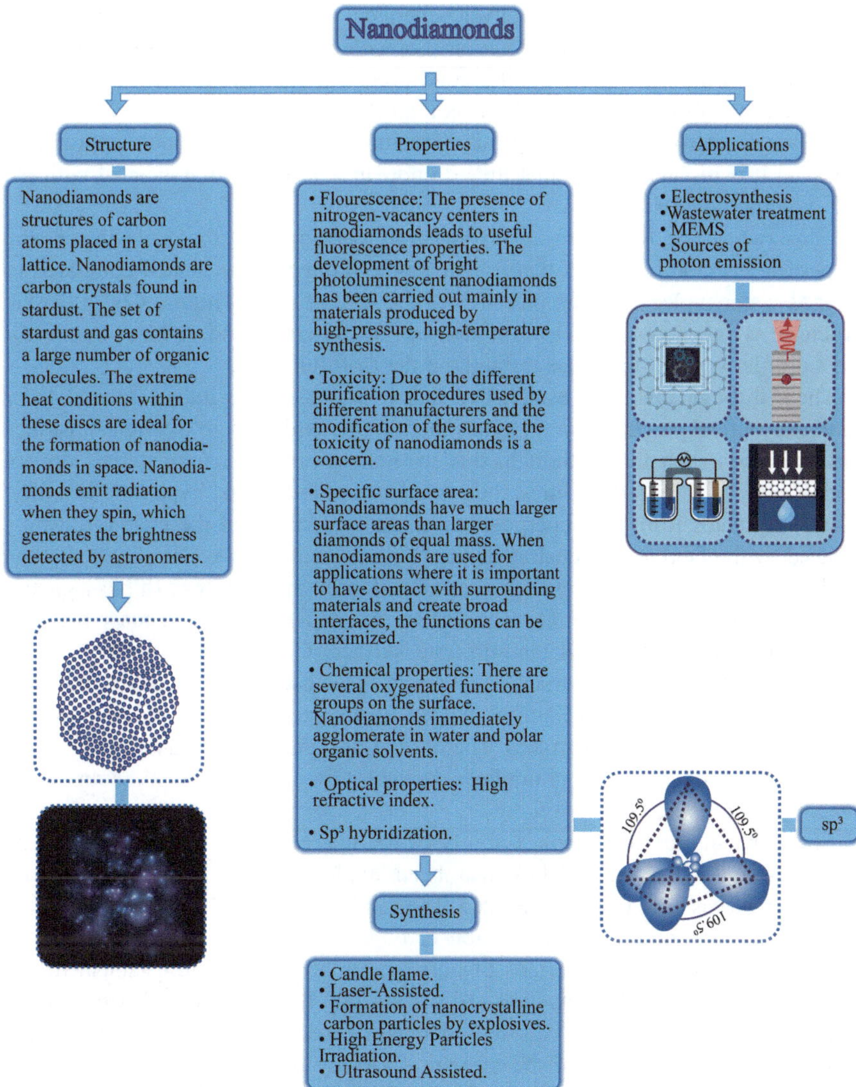

Fig. 3.8 Graphical synthesis of nanodiamonds properties and applications

3.2   **Fullerenes Second- and Third-Order Optical Nonlinearities**

As mentioned above, nonlinear optics is an important field due to the large number of applications in various fields. Many of the applications are based on organic nanomaterials in interfaces or substrates. The interaction between the materials before an optical excitation results in the modification of the optoelectronic properties of the nanomaterials due to the interface or substrate on which they are supported. So, understanding the physicochemical phenomena that affect interfaces is of utmost importance in order to control optoelectronic properties of devices and applications. Likewise, the study of second-order optical nonlinearities of materials supported on substrates and interfaces is important. On the other hand, C60 fullerenes have a centrosymmetric molecular structure, in the electric dipole approximation, and it is because of this that they do not show a response associated with the first hyperpolarizability, β.

Thus, when the fullerenes are synthesized with a water molecule inside, the properties are modified, such that a high dipole moment is presented compared to isolated water molecules. The presence of the water dipole within a C60 fullerene structure symmetrizes the electron density of fullerene and induces nonlinear second-order optical properties throughout the C60 fullerene.

Then, through this, it is established that it is possible to induce second-order nonlinear optical properties in materials that do not exhibit them through substrates and interfaces such that, although, due to their symmetry, nanomaterials do not exhibit early hyperpolarizabilities, their interface exhibits a first hyperpolarizability, which originates from the superposition of the materials and the charge transfers induced by the field of the substrate remains. Some studies have also shown that the nonlinear second-order optical response induced through interfaces exhibits a greater response associated with the interface surface and depends on the distance between the nanometric structure and the substrate surface. Consequently, in the case of centrosymmetric molecular structures, through the substrates or interfaces, they can present non-negligible contributions related to the second-order nonlinear optical response of functionalized surfaces. Therefore, this contribution is a consequence of the appearance of second-order nonlinear optical responses because of the rupture of the centrosymmetric [1].

Similarly, the first hyperpolarizability of the interaction between C60 fullerene structures and alkali metals has also been explored. So, C60 fullerenes with 12 sodium atoms have been synthesized. The results of this synthesis show nonlinear optical responses due to the interaction of both nanomaterials such that, as mentioned above, due to the interface of both, the nonlinearities are established. So, when alkali metal droplets are adsorbed on the surface of C60, large hyperpolarizability systems can be obtained. Likewise, the magnitude of the nonlinear optical response of the system is closely related to the way in which sodium atoms are adsorbed on the surface of fullerenes. Thus, when a large metallic droplet of sodium atoms is adsorbed at C60, large dipolar hyperpolarizabilities are expected. Likewise, through the dispersion of the sodium molecules, at various sites of the C60 fullerene structure, different adsorption sites are generated and large octupolar hyperpolarizabilities are also exhibited.

Then, another way to control nonlinear optical processes in nanomaterials is through the control of alkali metal adsorption in centrosymmetric systems, such that their hyperpolarizabilities can also be adjusted [2].

Structures of fullerenes with transition metals have also been studied. The second-order nonlinear optical response resulted due to the charge transfer transition between heteroatoms and carbon atoms in the carbon structure playing a fundamental role in the value of the first β hyperpolarizability. As a result, doping of the transition metal atom on fullerene may acquire a promising nonlinear optical response such that the analysis of nonlinear optical properties combined with the two-level expression explains why the substitution of a carbon atom with a hetero-atom in fullerene could improve the nonlinear optical response [3].

On the other hand, second-order nonlinear optical properties have been investigated by substituting -CH3 groups at different positions in a C60 fullerene structure. The second-order nonlinear optical responses of the C60 fullerene-fused dihydrocarboline derivatives can be adjusted by substitution of a strong electron acceptor donor. Studies on second-order nonlinear optical properties of dihydrocarboline derivatives in C60 fullerene structures have shown that the para-CH3 position is essential in the result of nonlinearities; in addition to the presence of a strong electron donor and acceptor in this position, we can adjust the nonlinear second-order optical response by altering the electronic properties of C60 [4].

Furthermore, the addition of Si in C58 and C59 fullerenes has been studied. This has also resulted in higher nonlinear second-order optical susceptibility, particularly for the $C58Si_2$ system. Fullerene molecules can be modified by adding silicon atoms to enhance their second-order photoinduced nonlinear optical properties. The results of the study through molecular dynamics geometry simulations show that the substitution of two silicon atoms favors the greater second-order susceptibility compared to the substitution of one atom. For the $C58Si_2$ molecule, the indicator of photoinduced second-order susceptibility is more isotropic compared to one atom substitution. Likewise, the results obtained for the microscopic hyperpolarizabilities using computer simulations for the C59Si and $C58Si_2$ molecules have a value very close to experimental data regarding the second generation of photoinduced optical harmonics [5].

3.3   Graphene Second- and Third-Order Optical Nonlinearities

A technique has been described to explore the second-order optical nonlinearities of graphene based on the strong fields per photon. Likewise, these fields are closely associated with confined graphene plasmons in combination with spatially nonlocal nonlinear optical interactions. The design of graphene nanostructures has been shown to allow the observation of strong internal nonlinearities. Likewise, the development of coupling techniques for individual nanostructures allows nonlinear processes to occur at the level of single-input photons. On the other hand, through the radiation of nanostructure matrices, internal nonlinearities are established that result in the frequency conversion of radiative fields at low input powers. On the other hand, nonlinearities are mostly observable in measurement systems that involve nanostructures in matrices. In this case, the light incident on the sample experiences frequency mixing at low input power due to interaction with plasmons. Likewise, individual nanostructures can generate nonclassical light states at low coupling efficiencies. Due to frequency coupling, second harmonic generation is possible [6].

As previously mentioned, the second-order nonlinear optical processes involve the mixing of electromagnetic waves for the generation of another optical frequency. On the other hand, although graphene constitutes a centrosymmetric structural medium, its second-order nonlinear optical response is different from zero if the effects of spatial scattering are considered. Thus, the possibility of a second-order nonlinear optical response of graphene has also been studied, including intraband and interband transitions, since the tensor associated with the resulting nonlinear optical susceptibility satisfies the symmetry properties necessary to carry out wave mixing processes [7].

Another way to break the centrosymmetric structures of graphene is with the application of external uniform electrostatic fields through hexagonal graphene quantum dots such that, through this technique, it is possible to induce the second-order nonlinear optical response in graphene quantum dots with a centrosymmetric structure. Since, through electrostatic fields, the geometric structure and electron density distribution of graphene quantum dots are broken, a first nonzero electronic hyperpolarizability results. The said induced electronic hyperpolarizability shows remarkable anisotropy for different directional fields such that the redistribution of electron density induced by a strong electric field can significantly reduce the orbital boundary energy of the quantum dot, improving the first electronic hyperpolarizability. The direction and strength of the field have an important influence on the magnitude of the first electronic hyperpolarizability. Thus, when the intensity of the electric field increases, likewise, the first hyperpolarizability increases, such that it reaches the maximum value of 162.37×10^{-50} C^3 m^3 J^{-2}. As it can be deduced, the strength and direction of the external electric field have an important influence on the geometric structures and second-order nonlinear optical properties of graphene quantum dots [8].

Besides, it has been shown that, regardless of the centrosymmetric structure of graphene, stacked bilayer graphene can exhibit a large and tunable second-order nonlinear susceptibility $\chi^{(2)}$. As seen above, by means of the application of an electric field, it is possible to induce nonlinear optical nonlinearities of the second order. Furthermore, nonlinear second-order optical susceptibilities $\chi^{(2)}$ have been shown to arise through quantum-enhanced photon processes due to the unique electronic spectrum of stacked bilayer graphene. Furthermore, the tunable electronic bandgap of the stacked bilayer graphene results in the tuning of the second-order nonlinear optical susceptibility resonance $\chi^{(2)}$. Thus, the magnitude of second-order nonlinear optical susceptibility $\chi^{(2)}$ is ~ 105 pm/V [9].

The result of the real and imaginary parts of the third-order nonlinear susceptibility $\chi^{(3)}$ has presented values of -8.9×10^{-10} esu and 3.6×10^{-8} esu, respectively, in the regime of fs [10]. Thus, it is possible to assume that the overlap between the graphene layers provides strong optical nonlinearities; in addition to the monolayer graphene, the nonlinearities can be associated with geometric disoriented stacks.

In addition, in the case of nonlinear third-order susceptibility, monolayer graphene presents a value of $\chi^{(3)} = 1.5 \times 10^{-7}$ esu in a region near infrared [11]. Correspondingly, the response of graphene through optical wave coupling is shown without dispersion in the wavelength range of 760–840 nm. At these wavelengths, a third-order nonlinear optical susceptibility $\chi^{(3)}$ response coincides in the regime of 1×10^{-7} esu [12].

The nonlinear optical susceptibility of graphene is several orders of magnitude higher than other dielectric nanostructured materials. Then, graphene can be used in imaging through the coupling of optical waves. Furthermore, the physical mechanism associated with nonlinearity can be attributed to the electromagnetic field that induces the reorientation of the atoms of the material. When incident light passes through a graphene layer, a transition between bands is generated which is a consequence of the excitation of the electron in the valence band due to the absorption of photons.

On the other hand, the third-order nonlinear optical response of the doped graphene presents a value of $\chi^{(3)} = 2.33 - 4.31 \times 10^{-11}$ esu [13]. Nonlinear susceptibilities exhibit dependencies on photon frequencies and the chemical potential derived from graphene doping. Thus, the chemical doping of doping allows the generation of desired nonlinearities by electrical tuning.

The nonlinear response of graphene is unaffected due to the wavelength resonance of the irradiated beams in the mixing of optical waves. However, in structures derived from graphene such as graphene quantum dot nanofluids, nonlinearities have occurred at different wavelengths, which is due to the geometry and optical characteristics of the material in which it is suspended. Suspensions based on graphene quantum dots have been mainly studied at wavelengths of 355 nm, 532 nm, and 1064 nm in the ps regime. Nonlinear optical response characteristics in such suspensions have been shown to be high at wavelengths close to the ultraviolet region. Then, the changes in the nonlinear refractive index due to the Kerr effect induced through the coupling of optical waves are quite significant at 355 nm, whereas the nonlinear optical response in the visible and infrared regions is not present [14].

Furthermore, the optical nonlinearities exhibited by graphene-covered semiconductor materials have a dependence on the distribution of the particles over energy states. So, optical nonlinearities have the potential to be electrically tunable. The nonlinear optical response of graphene-covered semiconductor materials at a wavelength of 1550 nm has exhibited remarkable results. Consequently, the coupling of waves on the material has shown that the nonlinear conductivity of graphene has a sharp resonance as a function of the beam irradiated on the sample. Besides, the geometry of the asymmetric resonance favors the absorption of the samples [15].

3.4 Graphene Oxide Second- and Third-Order Optical Nonlinearities

Graphene oxide shares a similar structure to graphene. The difference is that graphene oxide contains intercalated functional groups that contain oxygen. In general, it is a reduced form of graphene. The synthesis of graphene oxide consists of the oxidation and exfoliation of graphite.

Another way to generate breaks in the centrosymmetric structure of graphene is through dopants or structure variations, in addition to the placement of the nanostructures on a substrate or interface. Thus, through them it is possible to generate the second harmonic in groups that contain chemically reactive oxygen, such as graphene oxide, since oxygen establishes itself as a doping group that breaks the inversion symmetry in graphene. Likewise, the oxidation of nanomaterials leads to the interruption of the sp^2 domain, which also induces the opening of the bandgap that transforms nonluminescent graphene into luminescent graphene oxide. Thus, the value of the first hyperpolarizability of graphene oxide 2 sheets characterized in mass dispersed in an aqueous suspension, in the femtosecond regime at 800 nm of laser excitation, has exhibited a magnitude of 1.36×10^{-25} esu. This magnitude has been determined through the concentration-dependent harmonic dispersion signal. Also, these graphene oxide samples have exhibited multiple photon-excited fluorescence spectrum in a broad band in the visible range between 350 and 700 nm [16].

One of the main drawbacks of the measurement of molecular hyperpolarizabilities is that it cannot be deduced through direct observation of the second harmonic generation because the first hyperpolarizabilities are microscopic values. Macroscopic susceptibilities are related to molecular quantities through local field and molecular number density corrections. On the other hand, some factors that influence the measurement of hyperpolarizability are environmental interactions, such as solvent and other molecules [17].

Furthermore, the generation of controllable second harmonic through voltage has been studied at an interface of reduced graphene oxide with silicon. Likewise, the feasibility of controlling the generation of second harmonic tuning and the nonlinearity of a noncentrifugal system has been reported. On the other hand, the generation of adjustable second harmonic through voltage in graphene-based interfaces is of importance since it establishes the possibility of nonlinear photonic devices for flux applications and light interactions in complex systems. The design of these devices requires an understanding of the processes and parameters involved in the control of second-order nonlinear optical effects [18].

At the time, since they share a similar structure to graphene, this material also has a remarkable third-order nonlinear optical response. Also, graphene oxide consists of thin sheets of graphene, with a disordered structure. The third-order nonlinear optical response associated with graphene exhibits a value of $\chi^{(3)} = 2 \times 10^{-6}$ esu. While the reduced graphene oxide response exhibits a value of $\chi^{(3)} = 2.46 - 3.45 \times 10^{-6}$ esu [19].

As can be seen, graphene oxide, like reduced graphene oxide, exhibits optical responses that allow the material to be used as optical limiters. In addition, graphene oxide due to its structure allows it to be decorated with elements such as zinc ferrite. Thus, the decoration process allows the improvement of its nonlinear optical characteristics. The improvement exhibited in its nonlinear response can be attributed to the effect derived from the observed inverse saturable absorption and autofocus in the material.

The reduced graphene oxide decorated with zinc ferrite under at a wavelength of 532 nm and ultrafast excitation of 150 fs through the coupling of optical waves exhibited a value of $\chi^{(3)} = 4.2 \times 10^{-6}$ esu. On the other hand, at a wavelength of 800 nm and an ultrafast excitation of 150 fs at, a result of $\chi^{(3)} = 5.2 \times 10^{-15}$ esu was exhibited [20]. Third-order optical nonlinearity under wave laser excitations at 532 nm and 800 nm under a pulse regime of 150 fs shows notable differences. The enhanced nonlinear absorption arises mainly due to mechanisms related to the induced states that facilitate the transition between bands.

As pointed out, the reduced graphene oxide decorated with zinc ferrite exhibits a very good nonlinear optical response, which will allow the material to be used as a high-performance nonlinear photonic platform for nonlinear optical applications. On the other hand, unlike coated materials, it is possible to determine nonlinearities due to the Kerr effect of waveguides integrated with layers of graphene oxide. Through the coupling of waves at wavelengths of 1550 nm, an improvement in the nonlinear characteristics of hybrid waveguides is obtained, compared to semiconductor guides without graphene oxide layers. The above improvement is derived due to the influence of the wavelength and position of the optical wave coupling performance pattern [21].

3.5   Single-Wall Carbon Nanotubes Second- and Third-Order Optical Nonlinearities

As mentioned above, it is possible to generate non-centrosymmetric structures through doping of nanostructures. An example of this is the study of the dipole moment and the first hyperpolarizability of the single-walled carbon nanotube with nitrogen-doping effects at the edge of its structure such that the structure of single-walled carbon nanotubes with a higher number of doped nitrogen atoms has a higher dipole moment. On the other hand, the structure with a lower number of nitrogen atoms results in low excitation states. Thus, low values of optical hyperpolarizability occur. Likewise, the length of the single-walled carbon nanotube influences the hyperpolarizability response, such that elongation of the nitrogen-doped nanotube could amplify the first hyperpolarizability of the carbon nanotube. Thus, the first hyperpolarizabilities in nitrogen-doped single-walled super-short carbon nanotubes have been studied. Doping effects have been shown to greatly interfere with the nonlinear second-order optical response of carbon nanotubes such that, at a higher quantity of nitrogen, the magnitude of the first optical hyperpolarizability is almost 450 times greater than the hyperpolarizability of the undoped carbon nanotubes. Thus, it is shown that there is a relationship between the number of nitrogen doping and the value of optical hyperolarizability [22].

Likewise, the response derived from the second harmonic generation in single-walled carbon nanotubes is reported. To estimate the order of magnitude of second-order optical susceptibility $\chi^{(2)}$, a wavelength excitation at 800 nm was used. Also, the irradiated focal spot had a diameter of one-tenth the thickness of the single-walled carbon nanotube sample. On the other hand, the resonance peak was observed at 2 eV in the excitation spectrum of the second harmonic generation. Finally, as the main result, it was determined that the second-order optical susceptibility $\chi^{(2)}$ of the system based on single-walled carbon nanotubes has a magnitude of 2×10^{-6} esu. Thus, through these measurements it is established that second harmonic generation can be used to characterize the symmetry and chirality of single-walled carbon nanotubes [23].

Equally, the second harmonic generation in single-walled carbon nanotube films has been investigated through radiation at a wavelength of 1064 nm with an Nd: YAG laser source. This measurement has considered two classes of carbon nanotubes, with catalyst and without catalyst, of which only the sample synthesized without catalyst generated a second harmonic signal. This indicates that one of the factors that influence the second-order optical response is the catalyst used in the nanotube synthesis process [24].

On the other hand, it is also suggested that the structural non-centrosymmetry necessary for the second harmonic generation in carbon nanotubes is due to the nonzero chiral angle of single-walled carbon nanotubes. Therefore, the second harmonic optical generation can be observed from single-walled carbon nanotubes. Thus, the second harmonic generation of single-walled carbon nanotubes exhibited a magnitude of 9.1×10^{-28} W ms V^{-1} [25].

The study of the third-order nonlinear optical response in materials is useful to determine the functionality of materials in photonic applications. Single-walled carbon nanotubes attract attention due to their high third-order nonlinear optical response. This response is related to its tunable electronic transitions in the near infrared. Furthermore, unlike graphene, the optical response of nanotubes exhibits a resonant behavior as a function of the excitation wavelength. In the same way, the nonlinearities associated with single-walled carbon nanotubes are related to the structure of the nanomaterial. In relation to the above, it has been estimated that third-order susceptibility $\chi^{(3)}$ depends on the radius of the carbon nanotubes.

On the other hand, one of the most notable methods for measuring third-order nonlinear optical nonlinearities is optical wave mixing. From this method, it has been possible to obtain the value of third-order nonlinear optical susceptibility. The result exhibited by single-walled carbon nanotubes is $\chi^{(3)} = 3.3 \times 10^{-8}$ esu [26]. Thus, the large, exhibited values of third-order susceptibility point to promising applications of carbon nanotubes in photonics and optical detection.

It is also important to estimate the nonlinear optical responses of colloids involving single-walled carbon nanotubes. Thus, depending on the solvents of the colloid, their contributions from nonlinearities can be taken into account. Furthermore, the lengths of the carbon nanotubes are directly related to the nonlinear response. Then, the response of third-order optical nonlinearities has been reported using the optical Kerr effect in a regime of fs at a wavelength of 820 nm in solutions of single-walled carbon nanotubes. The third-order nonlinear optical susceptibility for solutions has shown a value of $\chi^{(3)} = 4 \times 10^{-13}$ esu [27]. This response is associated with the π-conjugated molecules of the nanotubes. Thus, it is also important to study the nonlinearities of single-walled carbon nanotubes at different lengths.

On the other hand, the third-order nonlinear susceptibility associated with thin films of single-walled carbon nanotubes is $\chi^{(3)} = 1 \times 10^{-8}$ esu through wave coupling [28]. In this sense, the mechanisms that promote the nonlinear response of the thin films of single-walled carbon nanotubes are linked to nonlinear absorption, nonlinear refraction, and scattering.

On the other hand, comparative studies of third-order optical nonlinearities $\chi^{(3)}$ exposed by carbon nanotubes have been performed in thick film samples at 532 and 355 nm. Carbon nanotubes under a wavelength of 532 nm and a regime of 10 ns, through the coupling of four optical waves, exhibited a value of $\chi^{(3)} = 4.5 \times 10^{-14}$ esu. On the other hand, at a wavelength of 355 nm, the result of nonlinearities was about six times greater than the value measured at 532 nm [29]. The results of the nonlinearities associated with the 355 nm wavelength of the carbon nanotubes suggest that there is a resonant nature.

Thus, the nonlinear response of single-walled nanotubes has an electronic contribution under resonant conditions. Also, the high values of the optical nonlinearities can be presented as a model that describes the phenomena and the order of magnitude of the nonlinearity.

3.6   Multiwall Carbon Nanotubes Second- and Third-Order Optical Nonlinearities

Multiwalled carbon nanotubes have a null value associated with the first optical hyperpolarizability due to their highly symmetric structure. However, through multiwalled carbon nanotubes with oligomer composite structures, it is possible to measure the coefficients associated with the response of second-order optical nonlinearities of multiwalled carbon nanotubes. For sample preparation, the oligomer was introduced longitudinally and transversely on the external surface of the multiwalled carbon nanotubes. Thus, it is due to the binding of a p-conjugated polymer that improved values of the first optical hyperpolarizability are exhibited. Then, the binding of the oligomer with the multiwalled carbon nanotubes results in potential nonlinear second-order response nanocomposites. Furthermore, the exceptional properties of the multiwalled carbon nanotubes associated with the oligomers contribute to the relevant properties of these hybrid materials. Thus, through these hybrid nanomaterials a panorama opens up for the design of applications in the field of optical nonlinearities because the values obtained from the first hyperpolarizability are very large compared to other nonlinear materials [30].

In addition, returning to the topic of colloids, another interesting factor to take into account when measuring the nonlinear response is the concentration of multiwall carbon nanotubes. Thus, in studies performed based on the concentration of nanotubes in suspensions in a ps regimen, the result of third-order susceptibility was $\chi^{(3)} = 2.8 \times 10^{-13}$ esu at a concentration of 0.4 mg/ml. Whereas, at a concentration of 1.6 mg/ml, the susceptibility is $\chi^{(3)} = 4.60 \times 10^{-13}$ esu. On the other hand, in a regimen of ns, the response at a concentration of 0.4 mg/ml is $\chi^{(3)} = 2.10 \times 10^{-12}$ esu. Also, at concentrations of 1.6 mg/ml it is $\chi^{(3)} = 3.43 \times 10^{-12}$ esu [31].

Also, the wavelength represents an important factor in the study of the nonlinearities of carbon nanotubes. Through the mixing of optical waves, the third-order optical susceptibility has been mainly explored in the regime of ps and ns at wavelengths, at 1064–532 nm. The result of third-order susceptibility at 1064 nm was $\chi^{(3)} = 6.4 \times 10^{-14}$ esu in ps regime, and $\chi^{(3)} = 1.17 \times 10^{-14}$ esu in ns regime. Also, the result of third-order susceptibility at 532 nm was $\chi^{(3)} = 6.3 \times 10^{-14}$ esu in ps regime, and $\chi^{(3)} = 0.39 \times 10^{-14}$ esu in ns regime [32]. Besides, through the coupling of optical waves it is possible to determine changes in the mechanical and physical properties of nanomaterials due to the nonlinearities exhibited by them, such as Young's modulus, density, and thermal conductivity.

In addition, the geometry of the multiwall carbon nanotubes allows them to be decorated with metallic particles. The purpose of this technique is to improve the optical characteristics of nanostructured materials such that the improvement of nonlinear optical response of multiwall nanotubes, due to the decoration of metallic nanoparticles, has been attractive.

Such is the case of the study carried out with a Kerr signal induced through the mixing of optical waves at a wavelength of 532 nm with pulses of ns. It is estimated that there is a significant change in third-order nonlinear optical susceptibility, so metallic nanoparticles play an important role in the optical properties of materials as observed. The third-order optical susceptibility response of multiwall carbon nanotubes is $\chi^{(3)} = 2.3 \times 10^{-9}$ esu. Whereas the third-order optical susceptibility response of decorated multiwall carbon nanotubes is $\chi^{(3)} = 1.3 \times 10^{-9}$ esu [33].

3.7 Highlights of Optical Nonlinearities in Carbon Nanostructures

It can be noted that carbon nanostructures offer large magnitudes of third-order nonlinear optical response. It is the case of carbon nanotubes, which have a nonlinear response associated with tunable electronic transitions in the near-infrared region. Equally, the nonlinearities of carbon nanotubes can exhibit a resonant behavior as a function of the incident wavelength. Another notable advantage is the structure of the carbon nanotubes. It is possible to control its third-order optical nonlinearities through the radius of the tubes. In the same way, the geometry of the carbon nanotubes allows decorations with metallic nanoparticles. Thus, through this technique it is possible to improve the optical characteristics of carbon nanostructured materials. On the other hand, the nonlinearities of graphene are associated with the transitions of electrons between bands of the material. Consequently, the increase in the third-order nonlinear optical response of graphene can result from mechanisms derived from the mixing of optical waves. The above is to establish contributions between coupling between layers of graphene to the resulting nonlinearities. Furthermore, the nonlinear optical susceptibility of graphene has been shown to be several orders of magnitude higher than that of other dielectric nanostructured materials. On the other hand, it is estimated that there is a significant change in third-order nonlinear optical susceptibility, which is due to metallic decorations on carbon nanostructures. Therefore, it can be stated that metallic nanoparticles play an important role in the optical properties of materials.

References

1. Nénon, S., & Champagne, B. (2014). Origin of the surface-induced first hyperpolarizability in the C60/SiO2 system: SCC-DFTB insight. *The Journal of Physical Chemistry Letters, 5*(1), 149–153.
2. Karamanis, P., & Pouchan, C. (2012). Fullerene–C60 in contact with alkali metal clusters: Prototype nano-objects of enhanced first hyperpolarizabilities. *The Journal of Physical Chemistry C, 116*(21), 11808–11819.
3. Liu, S., Gao, F. W., Xu, H. L., & Su, Z. M. (2019). Transition metals doped fullerenes: Structures–NLO property relationships. *Molecular Physics, 117*(6), 705–711.
4. Ma, N., Lv, M., Liu, T., Song, M., Liu, Y., & Zhang, G. (2019). Second-order nonlinear optical properties of [60] fullerene-fused dihydrocarboline derivates: A theoretical study on switch effect. *Journal of Materials Chemistry C, 7*(42), 13052–13058.
5. Fuks, I., Kityk, I. V., Kasperczyk, J., Berdowski, J., & Schirmer, I. (2002). Second-order nonlinear optical effects in C59Si and C58Si2. *Chemical Physics Letters, 353*(1–2), 7–10.
6. Manzoni, M. T., Silveiro, I., de Abajo, F. G., & Chang, D. E. (2015). Second-order quantum nonlinear optical processes in single graphene nanostructures and arrays. *New Journal of Physics, 17*(8), 083031.
7. Wang, Y., Tokman, M., & Belyanin, A. (2016). Second-order nonlinear optical response of graphene. *Physical Review B, 94*(19), 195442.
8. Li, H., Yu, X., Shen, X., Tang, G., & Han, K. (2019). External electric field induced second-order nonlinear optical effects in hexagonal graphene quantum dots. *The Journal of Physical Chemistry C, 123*(32), 20020–20025.
9. Wu, S., Mao, L., Jones, A. M., Yao, W., Zhang, C., & Xu, X. (2012). Quantum-enhanced tunable second-order optical nonlinearity in bilayer graphene. *Nano Letters, 12*(4), 2032–2036.
10. Chen, W., Wang, G., Qin, S., Wang, C., Fang, J., Qi, J., & Chang, S. (2013). The nonlinear optical properties of coupling and decoupling graphene layers. *AIP Advances, 3*(4), 042123.
11. Liu, Z., Zhang, X., Yan, X., Chen, Y., & Tian, J. (2012). Nonlinear optical properties of graphene-based materials. *Chinese Science Bulletin, 57*(23), 2971–2982.
12. Hendry, E., Hale, P. J., Moger, J., Savchenko, A. K., & Mikhailov, S. A. (2010). Coherent nonlinear optical response of graphene. *Physical Review Letters, 105*(9), 097401.

13. Cheng, J. L., Vermeulen, N., & Sipe, J. E. (2014). Third-order optical nonlinearity of graphene. *New Journal of Physics, 16*(5), 053014.
14. Wang, H., Ciret, C., Cassagne, C., & Boudebs, G. (2019). Measurement of the third order optical nonlinearities of graphene quantum dots in water at 355 nm, 532 nm and 1064 nm. *Optical Materials Express, 9*(2), 339–351.
15. Alexander, K., Savostianova, N. A., Mikhailov, S. A., Kuyken, B., & Van Thourhout, D. (2017). Electrically tunable optical nonlinearities in graphene-covered SiN waveguides characterized by four-wave mixing. *ACS Photonics, 4*(12), 3039–3044.
16. Russier-Antoine, I., Fakhouri, H., Basu, S., Bertorelle, F., Dugourd, P., Brevet, P. F., & Antoine, R. (2020). Second harmonic scattering from mass characterized 2D graphene oxide sheets. *Chemical Communications, 56*(27), 3859–3862.
17. Junqueira, G. M. A., Mendonça, J. P. A., Lima, A. H., Quirino, W. G., & Sato, F. (2016). Enhancement of nonlinear optical properties of graphene oxide-based structures: Push–pull models. *RSC Advances, 6*(97), 94437–94450.
18. Fernandes, G. E., Kim, J. H., Osgood, R., III, & Xu, J. (2018). Field-controllable second harmonic generation at a graphene oxide heterointerface. *Nanotechnology, 29*(10), 105201.
19. Muruganandi, G., Saravanan, M., Vinitha, G., Raj, M. J., & Girisun, T. S. (2017). Effect of reducing agents in tuning the third-order optical nonlinearity and optical limiting behavior of reduced graphene oxide. *Chemical Physics, 488*, 55–61.
20. Saravanan, M., Girisun, T. S., Vinitha, G., & Rao, S. V. (2016). Improved third-order optical nonlinearity and optical limiting behaviour of (nanospindle and nanosphere) zinc ferrite decorated reduced graphene oxide under continuous and ultrafast laser excitation. *RSC Advances, 6*(94), 91083–91092.
21. Qu, Y., Wu, J., Yang, Y., Zhang, Y., Liang, Y., El Dirani, H., & Monat, C. (2020). Enhanced four-wave mixing in silicon nitride waveguides integrated with 2D layered graphene oxide films. *Advanced Optical Materials, 8*(23), 2001048.
22. Xu, H. L., Wang, F. F., Li, Z. R., Wang, B. Q., Wu, D., Chen, W., & Aoki, Y. (2009). The nitrogen edge-doped effect on the static first hyperpolarizability of the supershort single-walled carbon nanotube. *Journal of Computational Chemistry, 30*(7), 1128–1134.
23. Su, H. M., Ye, J. T., Tang, Z. K., & Wong, K. S. (2008). Resonant second-harmonic generation in monosized and aligned single-walled carbon nanotubes. *Physical Review B, 77*(12), 125428.
24. De Dominicis, L., Botti, S., Asilyan, L. S., Ciardi, R., Fantoni, R., Terranova, M. L., & Appolloni, R. (2004). Second-and third-harmonic generation in single-walled carbon nanotubes at nanosecond time scale. *Applied Physics Letters, 85*(8), 1418–1420.
25. Huttunen, M. J., Herranen, O., Johansson, A., Jiang, H., Mudimela, P. R., Myllyperkiö, P., & Pettersson, M. (2013). Measurement of optical second-harmonic generation from an individual single-walled carbon nanotube. *New Journal of Physics, 15*(8), 083043.
26. Zhao, W. (2011). Four-wave mixing measurement of third-order nonlinear susceptibilities of length-sorted single-walled carbon nanotubes. *The Journal of Physical Chemistry Letters, 2*(5), 482–487.
27. Wang, S., Huang, W., Yang, H., Gong, Q., Shi, Z., Zhou, X., & Gu, Z. (2000). Large and ultrafast third-order optical nonlinearity of single-wall carbon nanotubes at 820 nm. *Chemical Physics Letters, 320*(5–6), 411–414.
28. Seo, J., Ma, S., Yang, Q., Creekmore, L., Battle, R., Tabibi, M., & Jung, S. (2006). Third-order optical nonlinearities of singlewall carbon nanotubes for nonlinear transmission limiting application. In *Journal of Physics: Conference Series*. IOP Publishing.
29. Ana, E. D. A., Dos Santos, H. F., & de Almeida, W. B. (2011). Enhanced nonlinearites of functionalized single wall carbon nanotubes with diethynylsilane derivatives. *Chemical Physics Letters, 514*(1–3), 134–140.

30. Botti, S., Ciardi, R., De Dominicis, L., Fantoni, R., Asilyan, L. S., Fiori, A., & Appolloni, R. (2003). DFWM: A proposed method to measure $\chi(3)$ on carbon nanotubes on the nanosecond time-scale. *Journal of Raman Spectroscopy, 34*(12), 1025–1029.
31. Jena, K. C., Bisht, P. B., Shaijumon, M. M., & Ramaprabhu, S. (2007). Study of optical nonlinearity of functionalized multi-wall carbon nanotubes by using degenerate four wave mixing and Z-scan techniques. *Optics Communications, 273*(1), 153–158.
32. Xie, S., Li, W., Pan, Z., Chang, B., & Sun, L. (2000). Mechanical and physical properties on carbon nanotube. *Journal of Physics and Chemistry of Solids, 61*(7), 1153–1158.
33. Mercado-Zúñiga, C., Torres-Torres, C., Trejo-Valdez, M., Torres-Martinez, R., Cervantes-Sodi, F., & Vargas-Garcia, J. R. (2014). Influence of silver decoration on the nonlinear optical absorption exhibited by multiwall carbon nanotubes. *Journal of Nanoparticle Research, 16*(4), 2334.

Chapter 4
Study on Second- and Third-Order Nonlinear Optical Properties in Metallic Nanoparticles

4.1 Metallic Nanostructures: Shape Classification and Properties

The propagation of electromagnetic modes through metallic nanoparticles has been studied due to the interaction of nanoscopic metals and light, which is a large field with interesting results. Metallic nanoparticles attract a great deal of attention due to their optical properties. The optical properties depend on the geometric properties of the metallic nanoparticles, such as the size and shape that arise from their resonances in the surface plasmon. Also, metallic materials are known as plasmonic materials. Surface plasmon resonances are unique to metals. Resonances are due to coherent oscillations of electrons of metals in a metal–dielectric interface as shown in Fig. 4.1.

Furthermore, the resonance frequencies of noble metals are in the middle of the electromagnetic spectrum, resulting in the strong interaction of surface plasmons with light. At the time, in particles smaller than the propagation length, the surface plasmons are limited to the geometry of the particle.

Then, in the case of metal nanoparticles, cavity resonators for surface plasmons can be contemplated. Thus, the surface plasmon resonances of the nanoparticles are known as localized surface plasmon resonances and depend on the size and shape of the metal nanoparticles. Notable interest in the study of plasmonic responses has developed due to new synthesis methods for a wide variety of metallic nanostructures. Similarly, it has increased nanoparticle characterization techniques, as well as numerical approximations for predicting the optical behavior of metallic nanostructures.

Noble metal nanoparticles have a unique localized surface plasmon resonance, which leads to their strong absorption and scattering as shown in Fig. 4.2. Thus, extinction spectra contain both absorption and scattering contributions that can serve multiple functions in different applications. The investigation of plasmonic responses has been of great importance since they determine the optical properties

© The Author(s), under exclusive license to Springer Nature Switzerland AG 2022

C. Torres-Torres, G. García-Beltrán, *Optical Nonlinearities in Nanostructured Systems*, Springer Tracts in Modern Physics 287, https://doi.org/10.1007/978-3-031-10824-2_4

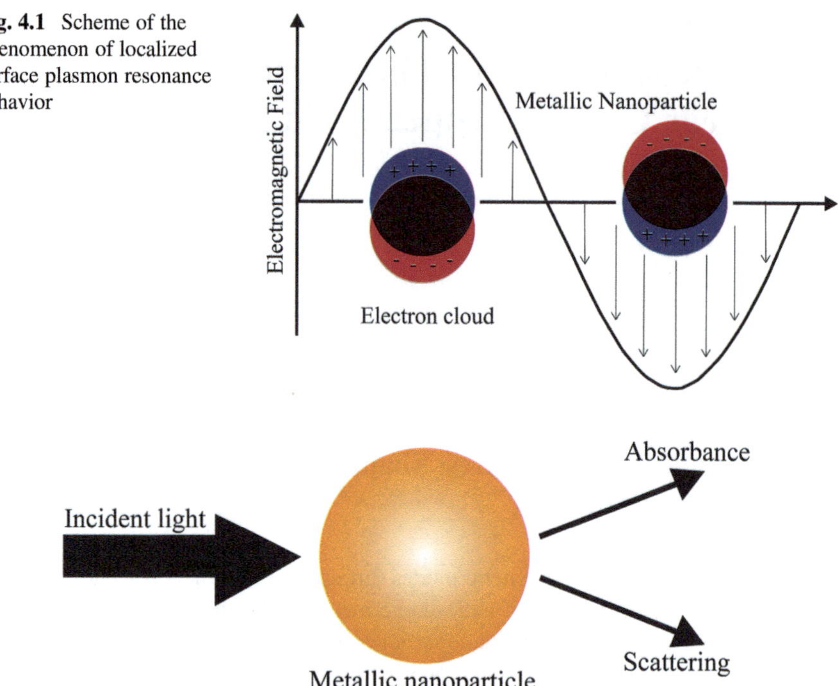

Fig. 4.1 Scheme of the phenomenon of localized surface plasmon resonance behavior

Fig. 4.2 Structure of extinction as a function of absorbance and scattering

and their potential applications in nanometric optical devices, sensors, photonic devices, and biosensors.

Therefore, to study the nonlinear optical response regarding metallic materials, it is necessary to differentiate the nonlinear contributions from the surface ones. It is therefore important to first define the optical contributions and then the surface resonances as described in Fig. 4.3.

Besides, the localized surface plasmon resonance also depends on the geometry and size of the nanostructures. Based on the number of dimensions on the nanometer scale, nanomaterials can be classified into controlled, 3D, multimetallic, and decorations. The different geometries of metallic nanostructures are described in Fig. 4.4.

Third-order nonlinearities in materials originate due to susceptibility $\chi^{(3)}$, which is defined as a complex function related to changes in the refractive index and absorption coefficient of materials. Besides, the nonlinear susceptibilities also depend on the frequency of the applied fields. In this way, it is possible to describe the susceptibilities as a function of polarization. Furthermore, the physical processes resulting from third-order polarization are closely related to the symmetry of the nonlinear material. Consequently, third-order nonlinear optical interactions can occur in nonlinear media with centrosymmetric and noncentrosymmetric geometries. Unlike second-order susceptibilities, they are exclusive to materials that do not show inversion symmetry, that is, they are not centrosymmetric. For this reason, in amorphous liquids, gases, and solids, the study of third-order nonlinear optical response is simple.

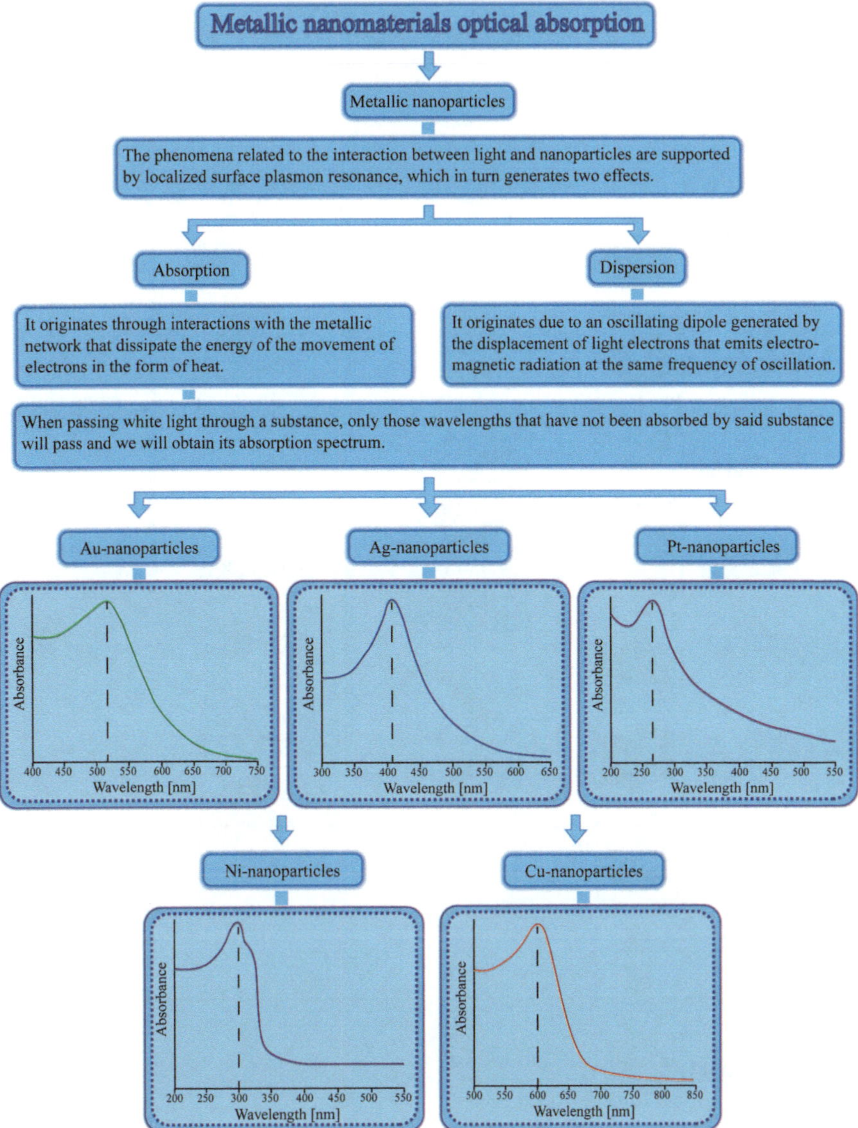

Fig. 4.3 Characteristic absorption spectra of metallic nanostructures

For the analysis of third-order susceptibilities, it is necessary to study the interactions considering the polarization of the waves. For the case of the study of nonlinear susceptibilities through the coupling of optical waves, it is possible to describe the polarization as a function of Fourier transforms. So, if the applied field is considered as the interaction of waves, it is necessary to consider the oscillation of

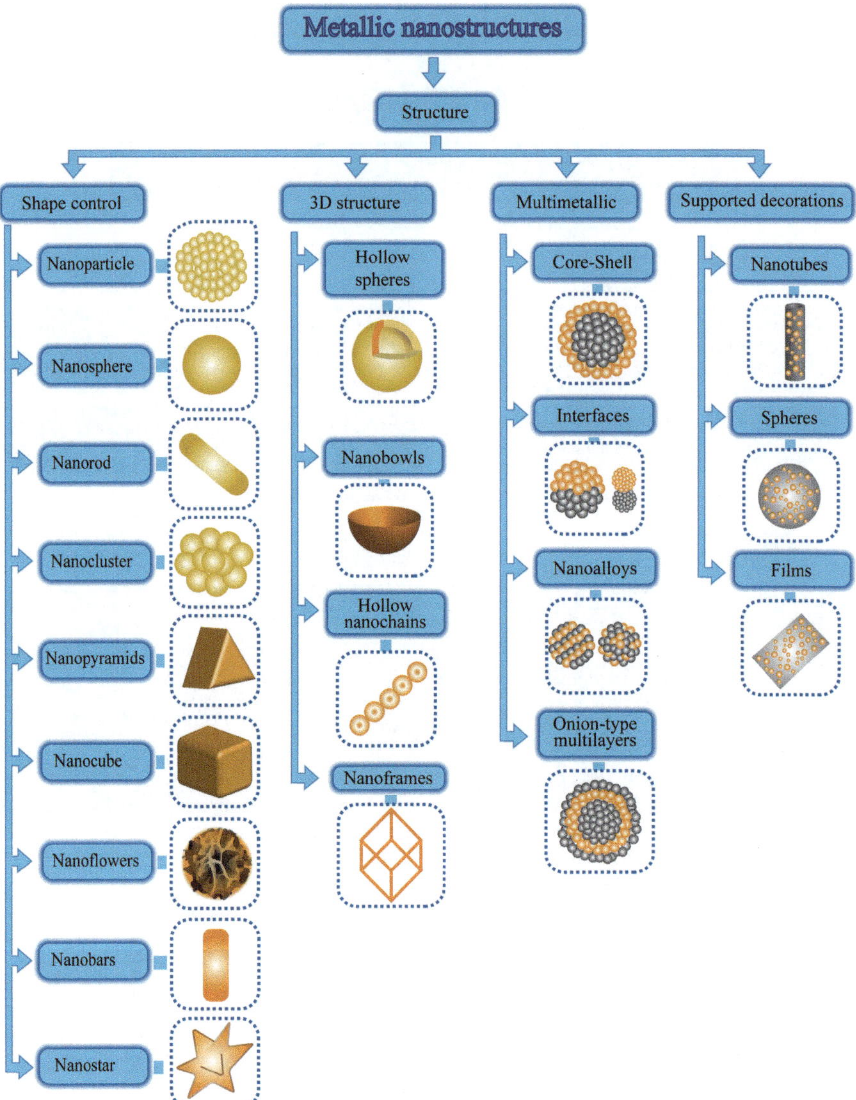

Fig. 4.4 Graphical illustration of nanostructured metallic shapes classification

the field frequencies, ω_1, ω_2, and ω_3. The polarization of the nonlinearities of the material expressed as a function of Fourier components of the nonlinear polarization is [1]

$$P_i^{(3)}(\omega_4) = \varepsilon_0 D^{(3)} \sum_{jkl} \chi_{ijkl}^{(3)}(-\omega_4; \omega_1, \omega_2, \omega_3) E_j(\omega_1) E_k(\omega_2) E_l(\omega_3) \qquad (4.1)$$

In addition, the nonlinear optical nonlinearity presented by metals is several orders of magnitude higher than in biological and organic samples. Furthermore, metallic nanoparticles present localized surface plasmons and surface plasmons, which increase the absorption coefficient of metallic materials.

Technological applications have been generated through the localized surface plasmon resonance exhibited by metallic nanoparticles. One of the applications mentioned is based on the detection of molecular interactions near the surface of the nanoparticle through changes in the spectral peak of the localized surface resonance plasmon of the material. Moreover, this phenomenon allows the optimization of the size of signal transduction sensors, in addition to biological detection sensors [2].

In addition, there are nanostructures that have the ability to present two localized surface plasmon resonances when interacting with a crystalline substrate. Given this behavior, it is possible to observe dielectric environmental dependencies for each resonance. Additionally, it has been observed that the morphology of the nanostructures greatly influences the spectroscopic response of the sample. Due to this, nanoparticles with spherical geometries do not exhibit double resonance behavior. However, if they are embedded in the surface of the substrate, they can exhibit double resonance behavior. This shows that the plasmon resonance structure of a nanoparticle in contact with a dielectric substrate is dependent on shape and size.

On the other hand, particles with cubic geometries are the most ideal for the reproduction of the double resonance phenomenon. This is due to the fact that cubic geometries induce large polarizations on the upper and lower surfaces of the particles. So, from one of the resonances you can develop detection applications due to its sharpness. Because of this, cubic nanostructures offer the possibility of designing applications such as chemical sensors. In addition, a research gap is opened in order to obtain a wider range of nanoparticle geometries with resonant properties [3].

Furthermore, it has been noted that there are improvements in the localized surface plasmon resonance response due to absorption and scattering by resonant plasmon nanoparticles. Moreover, numerical simulation from layer potential techniques is important to obtain a description of volume expansion for electromagnetic fields. Different simulations of individual nanoparticles and in group have predicted a nonlinear response with influence derived from the separation distance between nanoparticles.

Furthermore, the strong light scattering and enhancement of local electromagnetic fields are due to the localized surface plasmon excitation owing to the electromagnetic field at an incident wavelength that results. Metallic nanoparticles with localized surface plasmon resonances may contribute to biomedical applications. For instance, applications that include the detection of cancer cells and their photothermal ablation have been studied. In particular, the resonant plasmon nanoparticles provide good dispersion, absorption, and biocompatibility. In this direction, nanoparticles are suitable for use in therapeutic applications [4].

4.2 Gold Second- and Third-Order Optical Nonlinearities

The second harmonic generation described the generation of a photon based on the interaction of two photons, in addition to being one of the simplest nonlinear optical processes. However, because this method is feasible only for media that do not have inversion symmetry, this technique has been used on surfaces. So, second harmonic generation can be used to perform surface spectroscopy. In addition, it has been used to measure optical nonlinearities on flat metal surfaces. However, this technique can also be used in the measurement of metallic nanoparticles smaller than the wavelength of the light irradiated on the metallic sample. Measurements of gold nanoparticles covered with thioalkane have been made so that it is possible to measure the first hyperpolarizability due to the size of the nanoparticles where the electric dipole approximation is valid, and the higher-order contributions are sufficiently sensitive to the surface, such that the value of the first hyperolarizability for gold nanoparticles with thioalkane coating is 3.37×10^{-26} esu [5].

On the other hand, the first hyperpolarizability of metallic gold nanoparticles has been studied through the hyper-Rayleigh scattering technique. Through this technique, the second harmonic light derived from a liquid suspension of metallic gold particles is analyzed. Through this, it has been reported that the value of the first hyperpolarizability of gold nanoparticles, depending on their size, ranges from 0.5×10^{-25} esu, to 109×10^{-25} esu. The above optical response of second-order nonlinearities involves gold nanoparticles of sizes in the range of 10–150 nm. So, due to the dependence of the quadratic hyperpolarizability as a function of the particle size, it is inferred that the delay affects electromagnetic fields. Therefore, it is well known that there is a dependency on the frequency conversion process [6].

On the other hand, the nonlinear optical response of gold nanoclusters has been studied. A comparison has been made between the magnitudes of monodisperse Au15 nanoclusters atomically stabilized with glutathione molecules and protected monodisperse Au25 nanoclusters. The magnitudes of the first hyperpolarizabilities of the nanoclusters were obtained from the hyper-Rayleigh scattering technique. Thus, the first hyperpolarizability for the Au15 nanocluster samples equates to a magnitude of 509×10^{-30} esu. On the other hand, the samples associated with the Au25 nanoclusters exhibit a magnitude of the first hyperpolarizability corresponding to 128×10^{-30} esu. The difference between the magnitudes is associated with the difference in the number of gold atoms [7].

In other measurements, hyper-Rayleigh scattering has been used to measure second-order nonlinear optical behaviors in gold nanoparticle arrays from which, as a result, it is obtained that the value of the first hyperpolarizability of gold nanoparticle matrices is dependent on their size, whose magnitude ranges from 680×10^{-30} esu, to 3500×10^{-30} esu. The diameters associated with the gold nanoparticle samples are in the range of 5–10 nm. Through the measurement results, promising responses are observed. Furthermore, the large values of the nanoparticles are associated with the interaction of two resonance photons with the plasmon band of the colloids [8].

As has been observed, there is a great dependence between the diameter of the nanoparticles and the magnitude of the first optical hyperpolarizability. Derived from the above, a hyperpolarizability magnitude of 5.61×10^{-30} esu has been reported for a pure gold nanoparticle, whose diameter is 8 nm. Unlike the hyperpolarizability of a pure gold nanoparticle whose diameter is 9.5 nm, which has a magnitude of 6.07×10^{-30} esu [9].

One of the metals of greatest interest is gold due to its properties highlighted in Fig. 4.5, its plasmonic response, and high third-order nonlinear response. This derives a special interest from nonlinear optical interactions in plasmonic materials.

The value of the third-order susceptibility $\chi^{(3)}$ of gold due to the interaction of optical waves has been reported to be in the order of 1×10^{-11} esu [10]. Correspondingly, in the case of gold nanoparticles, it is possible to improve the response of third-order nonlinearities through the symmetry due to the coupling of gold nanoparticles. In the case of optical wave coupling, the third-order response of gold can increase up to four orders of magnitude. This is possible if a change in wavelength is generated. So, if the wavelength is shifted toward the infrared, the resonance of the localized plasmon of the metal shifts. In consequence, the coupled wave method to obtain the third-order nonlinear response is efficient in metallic nanoparticles because it is possible to control the distance of the particles [11].

On the other hand, the hybrid compounds of gold nanoparticles implanted in polymers, colloids, and films are also of peculiar interest due to the interactions of the materials. Furthermore, it is important to study the samples at different

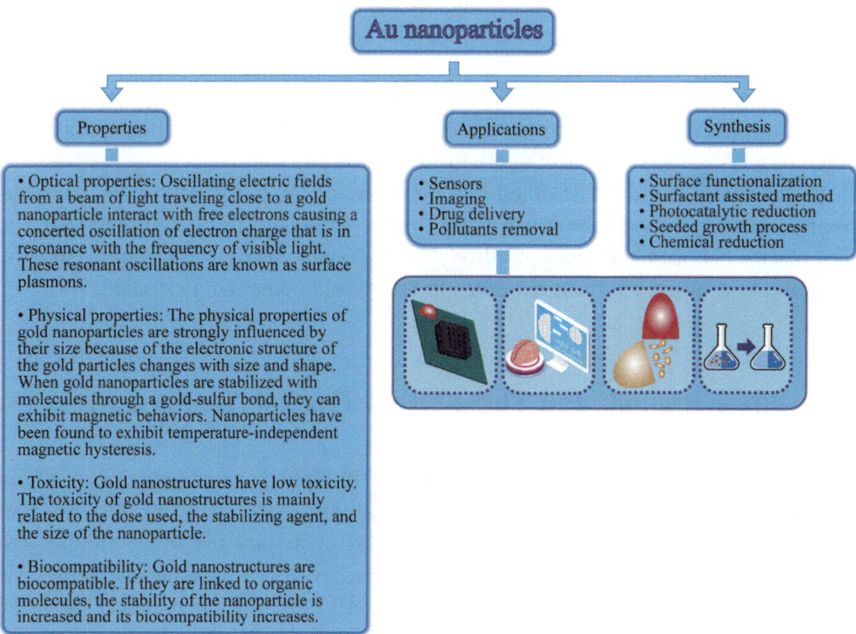

Fig. 4.5 Graphical synthesis of Au nanoparticle properties and applications

wavelengths, far from the resonance of the metallic nanoparticles. Also, the pulse rate of the beam that radiates the samples may have different physical mechanisms. Such is the case of ns pulses, in which the samples are subjected to thermal mechanisms responsible for the nonlinearities of the materials.

The physical phenomena that define the third-order susceptibility function are related to the real and imaginary components of $\chi^{(3)}$. So, the magnitude of third-order nonlinear optical susceptibility can be expressed as

$$\left|\chi^{(3)}\right| = \sqrt{\left(Re\,\chi^{(3)}\right)^2 + \left(Im\chi^{(3)}\right)^2} \tag{4.2}$$

The nonlinear refractive index n_2 and absorption coefficient β are parameters dependent on the different real and imaginary components of the third-order nonlinear optical susceptibility tensor.

For the Au nanoparticles in ion-implanted silica compounds, a third-order nonlinear optical response of $\chi^{(3)} = 2.2 \times 10^{-9}$ esu was presented. In addition to the real and imaginary components of the susceptibility function, they presented values of $Re\chi^{(3)} = -1.5 \times 10^{-9}$ esu and $Im\chi^{(3)} = -1.6 \times 10^{-9}$ esu [12].

Thus, gold hybrid materials exhibit third-order nonlinearity to optical wave coupling, based on the comparison of the light intensities of incident and self-diffracted polarized waves. Although they are of a lower order than the response exhibited by the coupled gold nanoparticles, they show considerable nonlinear optical response.

It is also possible that a variety of surface plasmon resonance peaks are generated through hybrid gold thin film compounds. The variation depends on the compounds of the films. In the case of Au/SiO$_2$, it presented resonances at wavelengths of 535 nm, while the composite films of Au/TiO$_2$ at 655 nm. For this sample, in studies related to optical coupled waves, the third-order optical susceptibility exhibited a value of $\chi^{(3)} = 2.6 \times 10^{-6}$ esu at a wavelength of 532 nm [13].

Furthermore, it should be noted that the value of the third-order optical susceptibility also depends on the pulse width of the irradiated beam. Furthermore, the differences between the susceptibility $\chi^{(3)}$ responses also closely depend on the concentration of the gold nanoparticles in the hybrid components studied.

4.3 Silver Second- and Third-Order Optical Nonlinearities

Measurements of silver nanoparticles covered with thioalkane have been carried out in order to measure the first hyperpolarizability associated with the second-order nonlinear optical response of the nanoparticles such that the value of the first hyperolarizability for silver nanoparticles with thioalkane coating is 2.10×10^{-26} esu for nanoparticles of 10 nm in diameter. The first hyperpolarizability of thioalkane-coated nanoparticles is found to be less than that of citrate-coated nanoparticles. This first reduced hyperpolarizability is tentatively attributed to the charge transfer that occurs between the silver atoms on the surface and the sulfur atoms in the coating layer [5].

In addition, using the hyper-Rayleigh dispersion technique, the second-order optical nonlinearities of suspensions of silver metallic nanoparticles, whose diameter varies between 20 nm and 80 nm, have been analyzed. As a result, the value of the first hyperpolarizability of silver nanoparticles, depending on their size, ranges from 4×10^{-25} esu to 194×10^{-25} esu. In addition, it is clear that the response obtained from the silver nanoparticles used in the measurements is of electrical dipolar origin and is manifested by the deviation of the shape of the nanoparticles with respect to that of a perfect sphere. Through this, it is established that the contributions associated with the electric dipole are derived from the size dependence of the values of the magnitude of the hyperpolarizability tensor with the volume of the particle. Furthermore, at a harmonic wavelength of 390 nm, enhancements in surface plasmon resonance occur [6].

Moreover, through hyper-Rayleigh scattering, the second-order nonlinearities of silver nanoparticles that arise from electric dipole and quadrupole plasmon resonances at the emitted wavelength have been analyzed; this nonlinear optical susceptibility is induced by the surface. From this, the magnitude of the first silver hyperpolarizability equates to a magnitude of approximately 5000×10^{-30} esu at an excitation wavelength of 820 nm. It should be noted that, at this wavelength, there is a contributing factor associated with the resonance between the nonlinear scattered radiation and the absorption of the silver dipole plasmon associated with a wavelength of approximately 400 nm [7].

In addition, it is very convenient to perform measurements of the hyper-Rayleigh scattering signals through suspensions with silver nanoparticles. Furthermore, through this it is possible to analyze the effect of the dispersion of the nanoparticles on the second-order nonlinear optical response of the metallic suspension. To this end, the suspension with nanoparticles of 10.5 nm in diameter was studied, and the response between samples with the addition of potassium nitrate and pyridine was compared. Through the above, it was observed that the hyper-Rayleigh signals improved at a higher concentration in the suspension. Also, a different improvement is expressed between the samples due to the complexity of the separated distances between the silver nanoparticles. Besides, the enhanced second-order nonlinearities are explained by the enhanced local electromagnetic field near the particle surfaces as they approach [8].

The improvement in the second-order optical nonlinearities in silver samples is also associated with the resonance in the scattered light in a nonlinear way with the plasmon absorption band of the surface of the particles. In addition, the second-order optical nonlinearities associated with silver nanoparticles are observed under two-photon resonance conditions [9].

The third-order optical response mechanisms of metallic silver nanoparticles depend on the physical and chemical properties mentioned in Fig. 4.6. This is in addition to the characteristics of the environment that surrounds it and its nonlinearities.

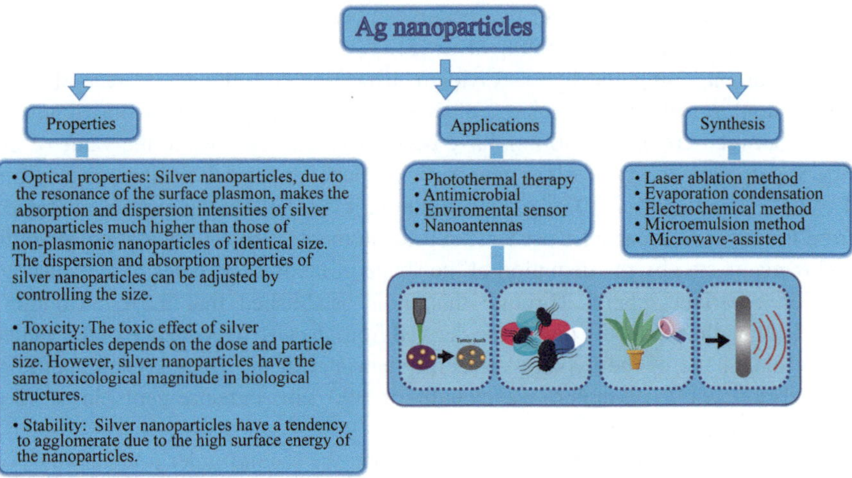

Fig. 4.6 Graphical synthesis of Ag nanoparticle properties and applications

It has been observed that the characteristics of the third-order nonlinear optical response associated with nanoparticles and metallic compounds are closely related to the wavelength irradiated on the material. Also, the resistance to optical damage associated with high-intensity laser excitation enables targeted applications in optical filters to be designed. In this way, it is also important to study the specific ablation threshold of metallic nanostructures and their relationship with their nonlinear optical response. The aforementioned technique is achieved through the exposure of Ag nanoparticles to an interaction of waves coupled to different laser irradiations of nanoseconds and picoseconds.

It has been observed that the characteristics of the third-order nonlinear optical response associated with nanoparticles and metallic compounds are closely related to the wavelength irradiated on the material. Also, the resistance to optical damage associated with high-intensity laser excitation enables targeted applications in optical filters to be designed. In this way, it is also important to study the specific ablation threshold of metallic nanostructures and their relationship with their nonlinear optical response. The study technique is achieved through the exposure of Ag nanoparticles to an interaction of waves coupled to different laser irradiations of nanoseconds and picoseconds.

As mentioned in the previous section, it is important to perform measurements at different wavelengths that imply the distance and closeness of the nanoparticle surface plasmon resonance. This is because the physical mechanisms presented at different wavelengths are different due to the excitation of the electrons of the material such that the nanoparticles irradiated near resonance exhibit intraband transitions in the ps regime and contribute energy distributions throughout the material. However, the energy distribution far from the resonance of the material exhibits an uneven energy distribution.

Then, the response exhibited by Ag nanoparticles at a wavelength of 355 nm close to the resonance of the material is $\chi^{(3)} = 3.33 \times 10^{-9}$ esu. Whereas the third-order susceptibility response at a wavelength of 532 nm has a value of $\chi^{(3)} = 2.13 \times 10^{-9}$ esu. Furthermore, in optical wave coupling, there is a higher self-diffraction efficiency at a wavelength of 532 nm compared to wavelength of 355 nm [14].

Other physical optical phenomena of materials are important to take into account in the study of third-order nonlinear responses, such as the scattering of light through materials. In the case of Ag nanoparticles embedded in silica glass, the dispersion of third-order optical susceptibility exhibits a value of $\chi^{(3)} = 6.54 \times 10^{-9}$ esu [15]. On the other hand, the dispersion of a third-order nonlinear optical susceptibility, for a Ag nanocrystal–glass composite obtained by coupled waves technique, is $\chi^{(3)} = 1.5 \times 10^{-10}$ esu [16].

Thus, the dispersion related to the third-order response of the Ag nanoparticles due to the localized surface resonance plasmon reflects an improvement of the internal electric field through the material. The above phenomenon is derived from the separation of the resonance and interband transitions.

4.4   Platinum Second- and Third-Order Optical Nonlinearities

The second-order nonlinearities of various solutions of metallic nanoparticles, including platinum, have been compared through the hyper-Rayleigh scattering technique. The platinum colloid was prepared from a mixture of nanoparticles with cubic and tetrahedral geometries averaging 8–10 nm in diameter. However, platinum nanoparticles do not show hyper-Rayleigh scattering signals detectable. This is associated with the fact that platinum nanoparticles lack plasmon absorption in the visible region [9].

Phthalocyanines are pigments made of a mostly organic molecule and are very convenient for creating hybrid materials with metallic nanoparticles because they are highly stable and stand out in their thermal resistance. Thus, by forming compounds with metallic nanoparticles, they are shown as potential materials for application in nonlinear optics as described in Fig. 4.7. In addition, it has third-order nonlinear optical susceptibility due to its plane geometry. In this way, the pigment molecules exhibit a third-order linearity, which varies according to the metal atom that is in the center of the said molecule.

It is convenient to measure the nonlinear optical susceptibility of platinum phthalocyanines because substitution of metals strongly enhances magnetic resonance imaging. Susceptibility can be measured through optical wave coupling. The nonlinear third-order susceptibility of platinum phthalocyanines at a wavelength of 1064 nm with pulses of 35 ps showed a value of $\chi^{(3)} = 2 \times 10^{-10}$ esu. Furthermore, studies affirm that the substitution of platinum in phthalocyanine pigments is more efficient than that of other metallic nanoparticles such as Pb [17]. Furthermore,

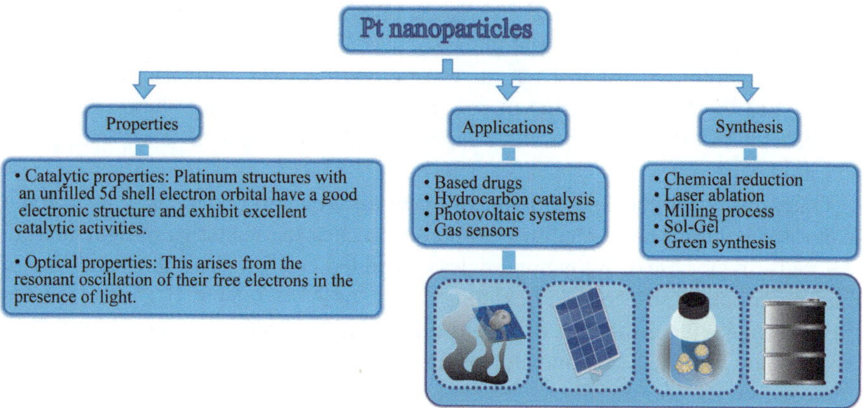

Fig. 4.7 Graphical synthesis of Pt nanoparticle properties and applications

metallic phthalocyanine pigments respond to physical mechanisms derived from the coupling of optical waves. They possess various low-intensity charge transfer states that are not present in the metal of free pigment. Thus, these states can contribute to increasing nonlinear susceptibility.

On the other hand, the optical response of the materials exhibits modifications due to thermal processes applied to the materials. Such is the case of samples of Pt ion-nanoparticles implanted in silica that through an external heat process modifies the nonlinear optical response exhibited by Pt nanoparticles. Furthermore, it is possible to modify the photoluminescent intensity, optical absorbance, and third-order nonlinear susceptibility. Thus, the modification in the parameters derived from the calorific mechanisms remains until a maximum correlated with the formation of the size of the metallic nanoparticles is achieved. Moreover, the plasmonic response of the Pt nanoparticles exhibits an important modification in the third-order properties associated with the ultraviolet range. Third-order optical susceptibility of Pt nanoparticles is exhibited in the range of 1×10^{-10} esu [18].

Therefore, it is important to explore the nonlinear optical response of Pt at infrared, visible, and ultraviolet wavelengths in order to expand the study also into the femtosecond, picosecond, and nanosecond regimes. In this way, it is possible to extend the possibilities of studying the different physical mechanisms in the interactions of waves with materials.

Third-order nonlinear optical properties exhibited by ion-implanted hybrid systems integrated by silicon quantum dots and Pt nanoclusters embedded in a silica exhibit a susceptibility of $\chi^{(3)} = 130 \times 10^{-11}$ esu due to Kerr gate mechanisms at 532 nm and 4 ns. Also, the mechanisms related to self-diffraction exhibit a susceptibility of $\chi^{(3)} = 7.2 \times 10^{-11}$ esu at 532 nm and 25 ps, in addition to $\chi^{(3)} = 9.5 \times 10^{-11}$ esu at 355 nm and 24 ps [19].

4.5   Nickel Second- and Third-Order Optical Nonlinearities

Chiral transition metal complexes are of great interest in the second-order nonlinear optical field due to their intrinsic non-centrosymmetric structure and the advantage of combining organic and inorganic compounds. Macromolecule complex with nickel nanostructures has been reported. Thus, the second-order nonlinear optical properties of the nickel macromolecule have been determined through time-dependent density functional theory. Furthermore, macromolecules and different substituent groups have been found to have a great effect on optical properties, such as electron absorption wavelengths, electron transition properties, and second-order nonlinear optical response. Some nickel macromolecules have a magnitude of the first hyperpolarizability of 32.6×10^{-30} esu [20].

As mentioned, there has been interest in the design of molecular materials because they have second-order nonlinear optical properties and a response of the first molecular hyperpolarizability. Nickel macromolecules are required to have strong low-energy electronic transitions and strong dipole moment variations during electronic excitation. These properties can be achieved thanks to the structural design of molecular compounds. These structures require characteristics such as an electron donor group and an electron receptor group bonded through a delocalized π system, which provides the polarizable electrons [21].

On the other hand, the second-order nonlinear optical properties of a nickel nanoparticle compound are evaluated by determining the electronic dipole moment, the first molecular hyperpolarizability. So, the magnitude of the first hyperpolarizability is 1.012×10^{-30} esu [22].

The study of materials that exhibit nonlinear optical properties has been far-reaching and far-reaching. Metallic materials have been shown to have a lot of potential related to third-order nonlinear optical response. These materials are related to important applications in the fields of optical communication, optical data storage, optical information processing, and quantum. One of the main solutions for improving third-order optical nonlinearities of materials has been based on the design of hybrid metallic materials. This approach is attractive due to the combination of organic molecules conjugated to transition metals.

A clear example of metallic hybrid materials is the three mixed ligand nickel (II) complexes. This compound exhibits a third-order nonlinearity outside resonance of $\chi^{(3)} = 3.20 \times 10^{-13}$ esu in the fs regime. The nonlinearities exhibited through wave coupling are related from the extensive delocalization and electronic polarization of Ni [23].

In the study of polymers, it has been deheating due to its low linear and nonlinear absorption. However, they are very good as base matrices to house metallic materials such as Ni. The nonlinear refractive index has been shown to be large for nickel dithiolene nanocomposites in a ps response regime.

The third-order optical nonlinearity response of a nickel dithiolene metal polymer through optical wave coupling exhibited very good results. The third-order susceptibility $\chi^{(3)}$ response presented a value in the range of $\times 10^{-17}$ esu in a ps regime [24]. The above result makes the exploration of the behavior of the third-order nonlinear optical response sustainable, as well as the absorption of two photons from systems of polymers and metallic dithiolenes.

4.6 Copper Second- and Third-Order Optical Nonlinearities

The nonlinear optical behavior of copper nanoparticles in aqueous solution has been examined through hyper-Rayleigh scattering. Copper nanoparticles prove to be excellent nonlinear dispersants because they have plasmon absorption bands on the surface of the intense visible region. Aqueous copper suspensions were prepared with 12-nm-diameter nanoparticles. Thus, the first optical hyperpolarizability associated with copper nanoparticles exhibited a magnitude of 1600×10^{-30} esu at an excitation wavelength of 820 nm [9].

Furthermore, through hyper-Rayleigh dispersion, the optical nonlinearities associated with copper nanoparticles in the size range of 5–100 nm have been studied. The measured hyperpolarizability values are associated with the particle size, which is important since it establishes that the second harmonic response of the copper nanoparticles results from the breaking of the symmetry on the surface of the particle instead of in the volume. The values of the first hyperpolarizability of copper nanoparticles have been studied at different wavelengths and nanoparticle diameters such that the values of the first hyperolarizability for copper nanoparticles at a wavelength of 1064 nm, range from $22{,}000 \times 10^{-30}$ esu, up to $23{,}830{,}000 \times 10^{-30}$ esu for nanoparticles with a range of 5–100 nm in diameter. On the other hand, at a wavelength of 1907 nm, the values of the first hyperolarizability range from 5200×10^{-30} esu, up to $1{,}500{,}000 \times 10^{-30}$ esu. Finally, at a wavelength of 1543 nm, the values of the first hyperolarizability range from $2{,}365{,}000 \times 10^{-30}$ esu, up to $1{,}500{,}000 \times 10^{-30}$ esu. It is then observed that there is a dependence between the size of the first hyperpolarizability of the copper nanoparticles at 1907 nm, and the hyper-Rayleigh scattering signal originates from the surface of the nanoparticle. However, the results of size-dependent measurements at 1064 nm establish a volume contribution [25].

On the other hand, in another study, the first hyperolarizability in nanoparticles was also measured through hyper-Rayleigh dispersion. These measurements were also carried out depending on the size of the nanoparticle in the range of 5–55 nm. The results of the measurements exhibited magnitudes ranging from $22{,}000 \times 10^{-30}$ esu, up to $5{,}200{,}000 \times 10^{-30}$ esu [26].

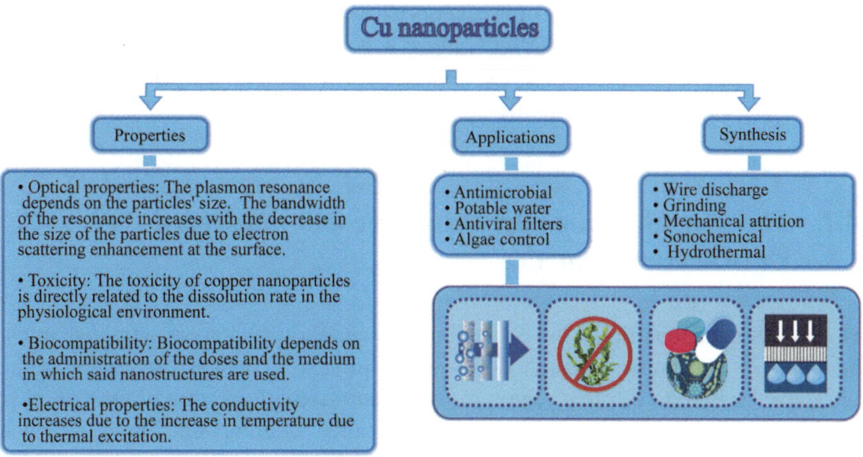

Fig. 4.8 Graphical synthesis of Cu nanoparticle properties and applications

Copper is a ductile metal with very high thermal and electrical conductivity as we can see in the description in Fig. 4.8. The morphology of copper nanoparticles is round, and their appearance is brown to black powder. Copper nanostructures are soft for some applications, so it is convenient to combine them with other metals to form stronger alloys. Also, copper nanoparticles are highly flammable and very toxic to aquatic life.

On the other hand, in the study of nanofluids, it is encouraging that the nonlinear responses exhibited are faster than the laser pulse duration of ps. This is important because they are strong candidates for nonlinear applications. Furthermore, metal nanoparticle nanofluids may exhibit improved nonlinear responses. So, it is important to discuss the mechanisms that contribute to third-order susceptibilities, and as mentioned above, one of them is plasmon resonance. It contributes to the improvements of the local field within the metallic particles. The third-order nonlinearities exhibited by glasses doped with copper nanoparticles also depend on the size of the metallic nanoparticles. Nonlinear susceptibility $\chi^{(3)}$ exhibits a maximum value of the order of 1×10^{-7} esu in a ps regime [27]. Besides, the result of the susceptibility $\chi^{(3)}$ response increases with the increase of the particles due to the dielectric constant of the metallic particles and the local field factor.

So, the need to design high-efficiency optical devices created the need to study hybrid optical materials embedded in other composites. Materials need to have high values in optical parameters such as nonlinear refractive index and nonlinear absorption coefficient. These parameters are associated with the third-order nonlinear optical susceptibility function. For this reason, there is considerable interest in nonlinear optics using metallic copper nanoparticles due to the high values resulting from nonlinear optical parameters and ultrafast response. In addition, they also have a response to the resonance plasmon. At the same time, it is possible to use copper nanoparticles in compounds such as $SrTiO_3$.

The results of the nonlinear optical properties of copper nanoparticles embedded in $SrTiO_3$ have been studied through optical wave coupling. The value of the third-order susceptibility exhibited by the mentioned compounds is in the range of $\chi^{(3)} = 1.55–5.33 \times 10^{-10}$ esu [28].

4.7 Multimetallic Nanostructures Second- and Third-Order Optical Nonlinearities

Noble metal nanoparticles have previously been shown to have attracted much attention in plasmonic research due to their improved optical properties in the ultraviolet visible to infrared region of the electromagnetic spectrum. On the other hand, nanoparticles of metallic alloys, also called multimetallic, have received attention due to the various combinations of alloys in order to obtain unique properties, electronic and optical. The phase miscibility property prevents metallic materials from mixing on a macro scale. However, in the nanometric regime the formation of bimetallic or multimetallic nanoparticle alloys is possible. These nanoparticles exhibit improved functionalities due to the effect caused by the mixture of elements. Furthermore, the advance in the synthesis of multimetallic nanostructures makes them good candidates for potential applications.

The most convenient metallic nanoparticles for designing plasmonic applications are based on the position of the resonance peak of the surface plasmon, in addition to its intensity and total width. Equally, it is possible to control the optical properties of the multimetallic alloy nanoparticles through geometric aspects; the size and the composition of each element are illustrated in Fig. 4.9.

On the other hand, in the case of coated multimetallic nanostructures, the optical properties can be adjusted through the control of core size and coating thickness. So, the surface plasmon resonance of the multimetallic nanoparticles can be adjusted in the large region of the electromagnetic spectrum by controlling the geometric factors. The high absorption in the visible region of the electromagnetic and near-infrared spectrum leads to possible photonic and image detection applications. In this manner, the multiple properties of multimetallic nanoparticles allow the design of applications in the field of biomedical, image detection, thermal phototherapy, drug delivery, catalysis, and contamination control.

The presence of free electrons in the conduction band of the surface of the metallic nanoparticles irradiated through a beam of light induces a collective excitation of these free electrons, as a consequence of the strong interaction with the

Fig. 4.9 Graphical images of structural models of multimetallic nanoparticles

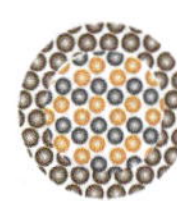
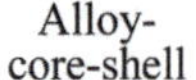

incident light. This results in the delocalized coherent oscillation of the confined electrons on the surface of the metallic nanoparticles at the metal–dielectric or metal–semiconductor interface. The electromagnetic excitation in the medium is called the polariton of the surface plasmon and is shown in Fig. 4.10. Polariton is a hybrid quasi-particle resulting from the strong coupling between light and free electrons on the surface of a nanoparticle.

Plasmonic nanoparticles can effectively absorb and scatter visible light, triggering localized surface plasmon resonance due to the interaction between electrons on the metal surface and incident light. Also, the strong transfer of plasmon resonance energy of metallic nanoparticles with biological samples is known as resonance energy transfer. In the interactions between metallic–biological nanoparticles, the gain medium can be described by electronic states with different energy levels. Thus, biological molecules can be extracted from a ground state S0 to excited states S1, ..., Sn as described in Fig. 4.11. Then, the resonant energy transfer can occur between the stimulated emission of the gain medium and the electromagnetic fields of the

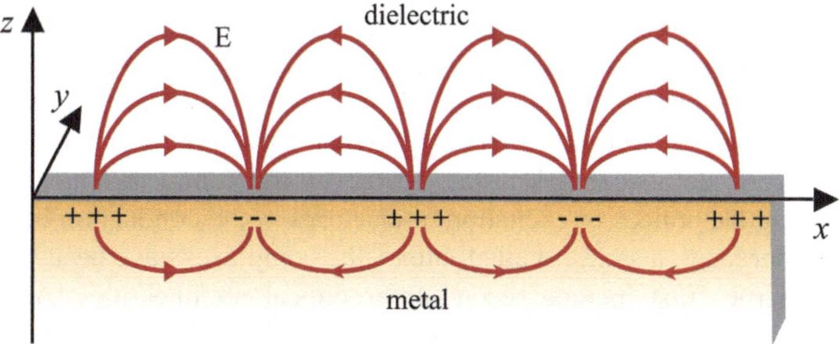

Fig. 4.10 Schematic diagram of polariton of the surface plasmon

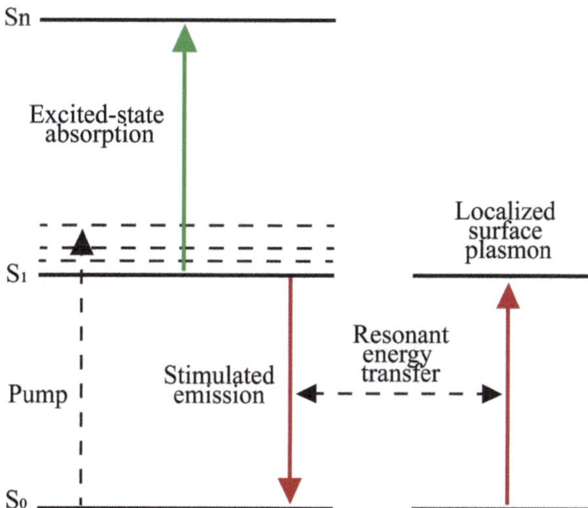

Fig. 4.11 Diagram of resonance energy transfer by metallic nanoparticles

surface plasmons derived from the metallic nanoparticles. Thus, the accumulation of surface plasmons in excited states can undergo stimulated emission. Then, through localized surface plasmon resonance interactions resulting from metallic nanoparticles, interaction with biological systems is possible, allowing a wide range of biomedical applications.

Multimetallic nanostructured materials attract attention because they outperform the physical and chemical properties of monometallic nanoparticles. Therefore, through the formation of joint nanometals, phenomena of synergistic effects are exhibited, and it is possible to establish a control of physicochemical properties through the change of morphology and concentration of multimetallic nanomaterials. Furthermore, nanostructured multimetallic materials modify their properties due to their size, as well as geometric and electronic effects.

The geometric effects are referred to the specific geometric orientation in relation to the number of atoms of the interacting metals on the surface of the multimetallic nanoparticle. On the other hand, the electronic effect refers to the modification of the electron density distribution as a consequence of the formation of mixed bonds. Besides, bimetallic nanoparticle synthesis methods are very convenient because growth control is feasible through the adjustment of the chemical reaction parameters. Similarly, metals are related to high thermal and electrical conductivity, as well as specific mechanical properties and high reflectivity of incident radiation. The exceptional properties of metals are related to their crystalline structure, in addition to the presence of delocalized electrons. Moreover, the nonlinear optical properties also depend on the collective excitation of electrons in the conduction band.

On the other hand, a series of alkali bimetallic complexes have been studied due to their properties that include second-order optical nonlinearities. Alkali metal adsorption is beneficial to the structure of bimetallic gold–germanium clusters. Furthermore, the alkali metals are distinguished by remaining strongly adhered to the surfaces or edges of the bunches Such that through these properties they are considerable with strong potential given by their second-order optical nonlinearities. The first hyperpolarizability is related to the cluster size and the geometric structure; it can be mentioned that bimetallic gold–germanium clusters exhibit optical hyperpolarizability values of up to $13{,}050 \times 10^{-30}$ esu [29].

Additionally, electrochemically coupled silver–gold bimetallic nanoclusters have been studied, with average sizes of 6 and 12 nm. Optical nonlinearities of these nanoparticles have been measured using the hyper-Rayleigh scattering technique. Through this, the first hyperpolarizability exhibiting a magnitude of 1.62×10^{-20} esu is estimated for coupled bimetallic nanoclusters with a size of 6 nm, which is four to five orders of magnitude higher than that of nanoparticles of gold [30].

The nonlinearities of aqueous solutions of gold and silver metal nanoparticle alloys, with a diameter between 15 and 50 nm, were evaluated with changes in concentration. Hyper-Rayleigh scattering measurements were performed to measure second-order optical nonlinearities of representative samples. The hyperpolarizability values obtained range from 400×10^{-22} esu, up to 200×10^{-22} esu. It has been observed that the hyperpolarizability of the metal alloy is large as the individual hyperpolarizability values for gold and silver nanoparticles. This is associated with the nonuniform distribution of metal atoms on the surface of the alloy [31].

The nonlinear optical third-order susceptibility of metallic Au–Ag nanoparticles has been measured at different wavelengths in the range from 700 nm to 950 nm. The results of the nonlinear parameter at 700 nm express $\chi^{(3)} = 5.30 \times 10^{-10}$ esu. At 750 nm, $\chi^{(3)} = 6.39 \times 10^{-10}$ esu. At 800 nm, $\chi^{(3)} = 7.42 \times 10^{-10}$ esu. At 850 nm, $\chi^{(3)} = 8.51 \times 10^{-10}$ esu. At 900 nm, $\chi^{(3)} = 9.57 \times 10^{-10}$ esu. At 950 nm, $\chi^{(3)} = 11.03 \times 10^{-10}$ esu [32].

Multimetallic nanostructured materials attract attention because they outperform the physical and chemical properties of monometallic nanoparticles. Also, through the formation of joint nanometals, phenomena of synergistic effects are exhibited, and it is possible to establish a control of physicochemical properties through the change of morphology and concentration of multimetallic nanomaterials. Similarly, nanostructured multimetallic materials modify their properties due to their size, as well as geometric and electronic effects.

The measurements of the nonlinear optical parameters of the Au–Ag nanoparticles express that derived from the resonance plasmon, the electromagnetic responses of the metal structure dominate in the order of wavelength plasmon resonance.

Similarly, in another study conducted for Au–Ag nanoparticles in solution, multimetallic nanoparticles exhibit characteristic photoelectron spectra and linear optical properties. The nonlinear optical properties of the multimetallic nanoparticles were measured in the absorption region of the surface plasmon resonance through the optical wave coupling technique. The third-order nonlinear optical susceptibility $\chi^{(3)}$ exhibited by Au–Ag nanoparticles in solution is of the order of 1×10^{-10} esu [33]. Thus, the third-order optical nonlinearity of Au–Ag multimetallic nanoparticles indicates that the nonlinear optical property does not depend on the metal species but depends mainly on the resonance effect of the surface plasmon.

Furthermore, Au–Ag multimetallic nanoparticles exhibit significant nonlinear optical response. Furthermore, nonlinearities are influenced by the position of the plasmon band in relation to the wavelength of excitation by the irradiated beam. The nonlinear optical response of the Au–Ag allied nanoparticles has been studied due to the increase in the Au molar fraction through laser excitation of wavelength 532 nm, 35 ps, and 4 ns. Third-order nonlinear susceptibility has been determined as a function of the mole fraction of Au–Ag nanoparticles contained in the samples. For samples with molar fraction Au = 0.8 and Ag = 0.2, $\chi^{(3)} = 18.3 \times 10^{-15}$ esu. To molar fraction Au = 0.6 and Ag = 0.4, $\chi^{(3)} = 7.7 \times 10^{-15}$ esu. To molar fraction Au = 0.5 and Ag = 0.5, $\chi^{(3)} = 6.3 \times 10^{-15}$ esu. To molar fraction Au = 0.4 and Ag = 0.6, $\chi^{(3)} = 4.2 \times 10^{-15}$ esu. To molar fraction Au = 0.2 and Ag = 0.8, $\chi^{(3)} = 3.2 \times 10^{-15}$ esu [34]. It is then observed that the increase in the Au molar fraction resulted in the displacement of the plasmonic band of multimetallic Au–Ag particles to a region close to the laser excitation wavelength, which causes an improvement in the resonance of the nonlinear optical response.

Equally, in another study it is expressed that the optical properties of the Au–Ag bimetallic nanoparticles exhibit a trend of third-order nonlinear optical susceptibility of a value $\chi^{(3)} = 9.34073 \times 10^{-8}$ esu [35]. For this reason, nanoparticles of Au–Ag and alloys have gained a lot of research attention due to their composition-dependent optical properties.

Otherwise, thin films with a thickness of 300 nm composed of Ag–BaO have exhibited nonlinearities due to the optical Kerr effect owing to the method of coupling optical waves at a wavelength of 820 nm. The third-order nonlinear optical susceptibility is exhibited by Ag–BaO $\chi^{(3)} = 4.8 \times 10^{-10}$ esu composite films [36]. The third-order optical nonlinearity due to the optical Kerr effect induced on the Ag–BaO thin films is attributed to the change in the refractive index due to the intraband transition of electrons from the occupied state near the Fermi level to the unoccupied state in the Ag nanoparticles.

Also, through the implementation of hybrid materials, it is possible to define multimetallic nanomaterials with magneto-optical properties, as is the case with multimetallic nanoparticles integrated by Fe–Ag, such that the third-order nonlinear optical response of Fe–Ag nanoparticles at a wavelength of 1550 nm is $\chi^{(3)} = 6.5 \times 10^{-12}$ esu [37].

The exceptional response of the nonlinearities of the magneto-optical samples is associated with the absorption of two photons. This phenomenon increases due to the synergistic effect induced by plasmons between the interacting metallic nanoparticles. Furthermore, nonlinear responses in Fe–Ag nanoparticle magneto-optical systems depend largely on the degree of overlap and, in turn, on the plasmon-induced dipole force through the metallic nanoparticles with ferromagnetic properties. Through this, it is established that the improvements in the third-order optical nonlinearities of the multimetallic Fe–Ag nanoparticles are due to the mechanisms corresponding to the resonance plasmons of the metallic materials.

On the other hand, the nonlinear optical parameters of Ag–CeO$_2$ multimetallic systems have been studied at a wavelength of 1064 nm with a pulse width of 50 ns. Third-order nonlinear optical susceptibilities indicate stronger results for Ag-doped CeO$_2$ samples compared to nondoped nanoparticles.

For samples of CeO$_2$ undoped, the third-order optical nonlinearities exhibit a magnitude of $\chi^{(3)} = 4.9 \times 10^{-6}$ esu. To samples of CeO$_2$ doped with Ag $= 0.25$ g, $\chi^{(3)} = 9.8 \times 10^{-6}$ esu. To CeO$_2$ doped with Ag $= 0.5$ g, $\chi^{(3)} = 10.7 \times 10^{-6}$ esu. To CeO$_2$ doped with Ag $= 0.75$ g, $\chi^{(3)} = 22.1 \times 10^{-6}$ esu. To CeO$_2$ doped with Ag $= 1$ g, $\chi^{(3)} = 19.4 \times 10^{-6}$ esu [38]. Remarkably, Ag-doped CeO$_2$ samples show an improvement of up to 4.5 greater than that of nondoped CeO$_2$.

Metal alloys containing Au or Ag nanoparticles in combination with other noble metals have been reported to exhibit improved electronic, optical, and mechanical properties. Moreover, Cu offers many similar properties to its noble metals, with the advantage that it is possible to improve the economic aspect; in addition, it is possible to easily generate alloys with other metallic nanoparticles. Due to this, the properties and synthesis of multimetallic alloy nanoparticles containing Cu with other metals are of interest.

On the other hand, the optical nonlinearities of glass systems have been the object of interest. For this reason, nonlinear optical properties of Cu–In alloys embedded in glass matrices have been investigated. Besides, the distribution and concentration of the doping of Cu_2In nanoparticles on the glass matrices are related to the nonlinear optical properties of the nanocomposites. The third-order nonlinear optical susceptibility for samples of Cu_2In doped with 0.5% mol exhibits a magnitude of $\chi^{(3)} = 1.51 \times 10^{-10}$ esu. Also, in samples of Cu_2In doped with 1.5% mol present a value of $\chi^{(3)} = 3.32 \times 10^{-10}$ esu [39]. The results show that the third-order nonlinear optical properties were improved with the increase in the doping concentration of Cu_2In nanoparticles, which have been up to three times higher than 0.5% by weight of embedded Cu_2In.

On the other hand, corresponding to the third-order nonlinear optical response of the Au–Cu samples, they have been determined in consideration of the heat exposure time of the multimetallic nanoparticle synthesis process. Then, the value of the third-order nonlinear susceptibility of an Au–Cu compound exposed for 10 h at a temperature of 450 ° C has a magnitude of $\chi^{(3)} = 3.1 \times 10^{-12}$ esu. For a sample exposed for 20 h, the optical response of third-order susceptibility exhibits $\chi^{(3)} = 5.4 \times 10^{-12}$ esu. Furthermore, the response for a sample exposed 30 h has a value of $\chi^{(3)} = 4.4 \times 10^{-12}$ esu [40]. The results of the optical nonlinearities have shown that the synthesis of Au–Cu nanocomposites on glasses has presented similar nonlinear behaviors with inverse saturation absorption effects and autodiffraction performance. Moreover, due to the chemical structure of the elements and the morphology of the glasses implanted with Au–Cu nanoparticles, the optical nonlinearities are related to the effects of quantum confinement and resonance of the surface plasmon.

Metal oxide nanoparticles can exhibit unique physical and chemical properties as shown in Fig. 4.12 due to their limited size and high density.

On the other hand, sodium borosilicate glasses embedded with Cu–Ni nanostructures have also been the object of study due to their nonlinear optical response. Furthermore, due to the synthesis process of multimetallic compounds, oxygen has been shown to produce randomly distributed CuO–NiO nanostructures in sodium borosilicate glasses. Whereas, due to the reduction of the atmosphere in the synthesis process, a Cu–Ni alloy nanostructure is generated in a glass host. However, both types of samples exhibit an effective reverse saturable absorption effect and autodiffraction performance. Thus, the third-order nonlinear optical susceptibility of the CuO–NiO multimetallic nanocomposites exhibits a magnitude of $\chi^{(3)} = 4.37 \times 10^{-14}$ esu. Equally, the third-order nonlinear optical susceptibility of the Cu–Ni multimetallic nanocomposites exhibits a magnitude of $\chi^{(3)} = 3.15 \times 10^{-12}$ esu [41]. The results report that the optical nonlinearities originate due to an electronic mechanism owing to the uniform distribution and the good dispersion of the Cu–Ni nanoparticles. Furthermore, due to the size of the distributed nanoparticles, the electronic contribution in the glass can be derived from the transition between bands. Also, thermal effects may contribute due to the repetition rate and pulse duration of fs. The nonlinear optical effect of the CuO–NiO multimetal nanocomposite is derived from the surface plasmon effect, while the

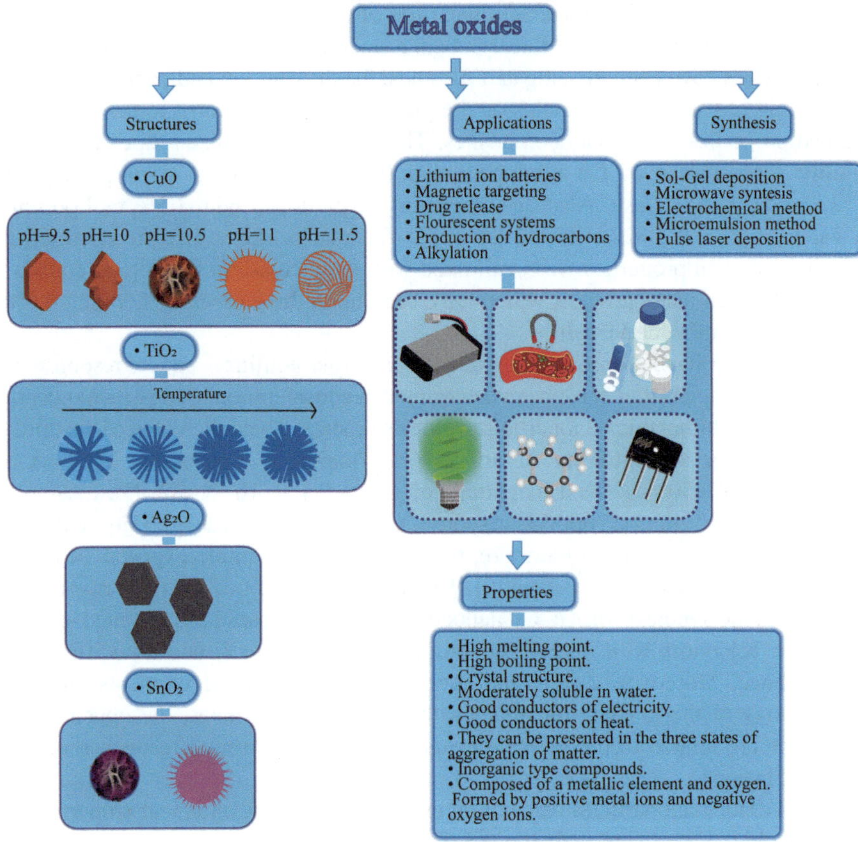

Fig. 4.12 Graphical synthesis of metal oxides nanoparticle properties and applications

nonlinearities of the Cu–Ni multimetal nanocomposite originate mainly from the electronic effect induced by the transition between bands.

On the other hand, the synthesis and nonlinear optical response of Au–Ni nanoparticle compounds embedded in sodium borosilicate glasses have also been studied. The synthesis of nanocomposites has made it possible to generate Au–NiO nanoparticles with random distributions in glasses for the purpose of conducting optical studies. Thus, the third-order nonlinear optical response for glass multimetallic nanocomposites with embedded Au–NiO nanoparticles exhibits a value of $\chi^{(3)} = 5.51 \times 10^{-12}$ esu [42]. The nonlinear third-order optical response of multimetallic nanocomposites demonstrates higher nonlinearities than those of pure Au nanoparticles. Furthermore, the third-order nonlinear optical response demonstrates an improvement due to the Au–NiO nanoparticles owing to the contribution of nonlinear refraction and absorption. This is due to the fact that NiO nanoparticles have negligible nonlinear absorption coefficients but a stronger nonlinear refractive response.

Also, the third-order nonlinear optical susceptibility measurement results at a wavelength of 800 nm indicate that Ni nanoparticle samples embedded in sodium borosilicate glasses have a magnitude of $\chi^{(3)} = 5.51 \times 10^{-12}$ esu [43]. It is also observed that the nonlinear response of improvement in the sodium borosilicate glasses originates from the Ni nanoparticles embedded in the glass matrix. It is particularly noteworthy since the results of third-order nonlinear optical susceptibility of sodium borosilicate nanocomposites doped with Ni nanoparticles are up to three orders of magnitude more favorable than that of pure sodium borosilicate, which exhibits a value of $\chi^{(3)} = 7.81 \times 10^{-15}$ esu.

Correspondingly, in another study, nonlinear optical susceptibility measurements were made of Cu–Ni nanoparticles contained in sodium borosilicate glasses synthesized through a sol–gel process. The third-order nonlinear optical properties of the glass doped with Cu–Ni nanoparticles were measured through a pulse in the fs regime. Derived from this, the value of third-order nonlinear optical susceptibility was calculated as a magnitude of $\chi^{(3)} = 4.92 \times 10^{-11}$ esu [44].

This is mainly related to the quantum confinement effects of Ni nanoparticles. Besides, the stability of the sodium borosilicate glass matrix is established through the uniform distribution and dispersion of the Ni nanoparticles, which is also associated with the high third-order optical nonlinearities. The results of the measurements also express that the progress of the nonlinearities of sodium borosilicate compounds doped with Ni nanoparticles results from the saturated absorption and the autofocus behavior of the nonlinear refraction. These results reveal that doped glass compounds possess the essential characteristics for potential application in nonlinear optical devices. Glasses containing metal nanoparticles have been shown to exhibit improved third-order nonlinear optical susceptibility. Corresponding to the third-order nonlinear optical response, the real part is related to the refractive index dependent on the irradiated beam intensity. Furthermore, optical properties are derived from the behavior of electronic systems, which, in turn, exhibit characteristics derived from dielectric effects and quantum confinement. Intraband and interband electronic transitions contribute to effective third-order nonlinear optical susceptibility. Furthermore, they turn out to depend on the type of metal and the shape and size of the nanoparticles, as well as the metal–dielectric bonds of the nanocomposites.

In the same way, the third-order nonlinear optical response exhibited by samples of TiO_2 matrices doped with Au–Pt bimetallic nanoparticles has been analyzed. The measurements of the nonlinear optical parameters have been carried out through the mixing of optical waves in order to describe the mechanisms derived from the optical phase change and the photothermic phenomena induced on the samples. In addition, through the optical wave coupling technique, it is possible to explore the automatic generation of polarized autodiffraction through the participation of the Kerr optical effect through multimetallic compounds. The third-order nonlinear optical susceptibility response of the TiO_2 nanocomposites doped with Au–Pt bimetallic nanoparticles studied at a wavelength of 532 nm exhibited a magnitude of $\chi^{(3)} = 4.6 \times 10^{-9}$ esu [45]. The corresponding measurements have made it possible to study the effect of self-diffraction induced due to the coupling of optical waves and in consideration of photothermal mechanisms.

In another way, the nonlinearities exhibited by Au–Pt multimetallic nanoparticles have been dynamically studied through ethanol suspensions. This procedure has also been carried out through an optical wave mixing process at 532 nm. In measurements, it has been identified that the optical Kerr effect is the main precursor of third-order nonlinear optical susceptibility. Furthermore, through dynamic measurements of multimetallic particles suspended in ethanol, the contribution of different surface plasmon resonances has been identified. This is due to the multiple excitation of the multimetallic nanoparticles. Thus, changes in the refractive index of the sample represent a change in third-order optical nonlinearities due to the dynamic rotation of the polarization of the beams interacting with the sample. The change in the refractive index due to third-order nonlinearities exhibited a representative value of $n_2 = 7.3 \times 10^{-16}$ m^2/W [46].

Therefore, Au–Pd samples have been studied through wavelengths of 514 nm. Nonlinearity measurements have been carried out at different concentrations of Au–Pd multimetallic nanoparticles. The nonlinearity properties of the Au–Pd nanofluid prepared at different concentrations have exhibited a maximum value of $n_2 = 13 \times 10^{-8}$ cm^2/W, and a value at a minimum concentration of $n_2 = 13 \times 10^{-8}$ cm^2/W [47]. The magnitude of the changes in the refractive indices of the samples due to nonlinearities shows a linear increase as the concentration ratio of the Au–Pd multimetallic nanoparticles increases. The results of the measurements determine that the behavior of the nonlinearities of Au–Pd multimetallic nanofluids has significant nonlinear refractive index values due to which they are considered for nonlinear thermal applications. Otherwise, combinations of palladium with organic elements have also been studied in the form of thin films of palladium oxide, which have shown significant differences from multimetallic nanocomposites. The third-order nonlinearities of thin films of palladium oxide have been analyzed through the optical wave coupling technique. It was determined that the third-order nonlinear susceptibility of palladium oxide films has a magnitude of $\chi^{(3)} = 1.6 \times 10^{-4}$ esu [48]. Moreover, it was observed that there is a thermal mechanism that causes the dynamic reduction of the refractive index derived from the interaction of beams with a wavelength of 532 nm. Regarding the possible nature of the high negative refractive nonlinearity of palladium oxide films, two effects are considered due to the interaction of the beams in the films. The first determines a few free carriers that are excited by a two-photon absorption process. In addition, there is a thermal mechanism that derives a short-wavelength shift around the absorption edge.

Plasmonic nanomaterials have been shown to offer promising optical approaches that can be used in different applications. Materials with plasmonic responses have been widely exploited because they exhibit a photon absorption mechanism in metallic materials. Besides, it is noteworthy that the plasmonic improvement of nanomaterials has been exposed through the development of hybrid nanostructures that integrate more than two different metallic elements. To know and adapt the control of the optics, mechanics, electronics, and other relevant properties of materials, it is necessary to study hybrid plasmonic nanostructures, such as multimetallic nanoparticles, grouped metallic nanoparticles, carbon–metal nanohybrids, and polymeric nanohybrids as shown in Fig. 4.13.

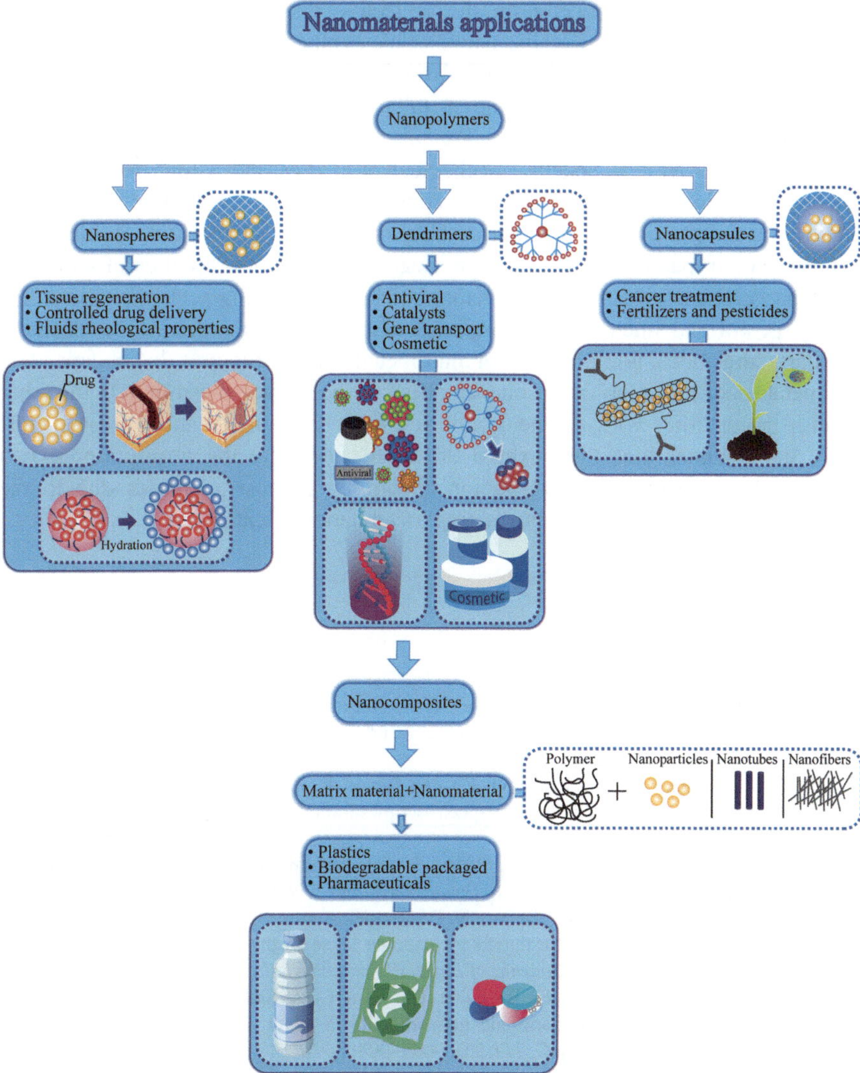

Fig. 4.13 Graphical synthesis of nanopolymer applications

4.8 Highlights of Optical Nonlinearities in Metallic Nanostructures

Multimetallic nanostructured materials attract the attention of optical studies because, through the formation of nanometals, synergistic effects phenomena are exhibited. From the above, it is possible to envision a control of third-order nonlinearities since nanostructured multimetallic materials modify their properties due to their size, as well as geometric and electronic effects. On the other hand, nonlinear optical properties also depend on the collective excitation of electrons in the conduction band. Also, bimetallic nanoparticle synthesis methods are very convenient because growth control is feasible by adjusting the parameters of the chemical reaction. Other advantages of using metallic nanostructures are related to their physicochemical properties. Metals are related to a high thermal and electrical conductivity, as well as specific mechanical properties and a high reflectivity of incident radiation. The properties are a consequence of its crystalline structure, in addition to the presence of delocalized electrons. On the other hand, multimetallic nanostructured materials exceed the physical and chemical properties of monometallic nanoparticles. Multimetallic nanoparticles exhibit characteristic photoelectronic spectra and linear optical properties derived from the resonance plasmon. Similarly, the optical nonlinearities of multimetallic nanostructures do not depend on the metal species, but on the resonance effect of the surface plasmon. On the other hand, nanomaterials with magneto-optic response have also been of great interest since the nonlinearities they exhibit are associated with the absorption of two photons. Furthermore, the two-photon absorption process increases due to the plasmon-induced effect of the interacting metal nanoparticles.

References

1. Sutherland, R. L. (2003). *Handbook of nonlinear optics*. CRC press.
2. Mayer, K. M., & Hafner, J. H. (2011). Localized surface plasmon resonance sensors. *Chemical Reviews, 111*(6), 3828–3857.
3. Sherry, L. J., Chang, S. H., Schatz, G. C., Van Duyne, R. P., Wiley, B. J., & Xia, Y. (2005). Localized surface plasmon resonance spectroscopy of single silver nanocubes. *Nano Letters, 5*(10), 2034–2038.
4. Ammari, H., Deng, Y., & Millien, P. (2016). Surface plasmon resonance of nanoparticles and applications in imaging. *Archive for Rational Mechanics and Analysis, 220*(1), 109–153.
5. El Harfouch, Y., Benichou, E., Bertorelle, F., Russier-Antoine, I., Jonin, C., Lascoux, N., & Brevet, P. F. (2012). Effect of a thioalkane capping layer on the first hyperpolarizabilities of gold and silver nanoparticles. *Journal of Physics: Condensed Matter, 24*(12), 124104.
6. Russier-Antoine, I., Benichou, E., Bachelier, G., Jonin, C., & Brevet, P. F. (2007). Multipolar contributions of the second harmonic generation from silver and gold nanoparticles. *The Journal of Physical Chemistry C, 111*(26), 9044–9048.
7. Hao, E. C., Schatz, G. C., Johnson, R. C., & Hupp, J. T. (2002). Hyper-Rayleigh scattering from silver nanoparticles. *The Journal of Chemical Physics, 117*(13), 5963–5966.
8. Wang, G., Zhang, Y., Cui, Y., Duan, M., & Liu, M. (2005). Study on the behavior of hyper-rayleigh scattering for silver nanoparticles with aggregation effects. *The Journal of Physical Chemistry B, 109*(3), 1067–1071.
9. Johnson, R. C., Li, J., Hupp, J. T., & Schatz, G. C. (2002). Hyper-Rayleigh scattering studies of silver, copper, and platinum nanoparticle suspensions. *Chemical Physics Letters, 356*(5–6), 534–540.
10. Boyd, R. W., Shi, Z., & De Leon, I. (2014). The third-order nonlinear optical susceptibility of gold. *Optics Communications, 326*, 74–79.
11. Danckwerts, M., & Novotny, L. (2007). Optical frequency mixing at coupled gold nanoparticles. *Physical Review Letters, 98*(2), 026104.
12. Torres-Torres, C., Khomenko, A. V., Cheang-Wong, J. C., Rodríguez-Fernández, L., Crespo-Sosa, A., & Oliver, A. (2007). Absorptive and refractive nonlinearities by four-wave mixing for Au nanoparticles in ion-implanted silica. *Optics Express, 15*(15), 9248–9253.

13. Liao, H., Lu, W., Yu, S., Wen, W., & Wong, G. K. (2005). Optical characteristics of gold nanoparticle-doped multilayer thin film. *Journal of the Optical Society of America B, 22*(9), 1923–1926.

14. Torres-Torres, C., Peréa-López, N., Reyes-Esqueda, J. A., Rodríguez-Fernández, L., Crespo-Sosa, A., Cheang-Wong, J. C., & Oliver, A. (2010). Ablation and optical third-order nonlinearities in Ag nanoparticles. *International Journal of Nanomedicine, 5*, 925.

15. Sato, R., Momida, H., Ohnuma, M., Sasase, M., Ohno, T., Kishimoto, N., & Takeda, Y. (2012). Experimental dispersion of the third order optical susceptibility of Ag nanoparticles. *Journal of the Optical Society of America B, 29*(9), 2410–2413.

16. Hamanaka, Y., Nakamura, A., Hayashi, N., & Omi, S. (2003). Dispersion curves of complex third-order optical susceptibilities around the surface plasmon resonance in Ag nanocrystal–glass composites. *Journal of the Optical Society of America B, 20*(6), 1227–1232.

17. Shirk, J. S., Lindle, J. R., Bartoli, F. J., Hoffman, C. A., Kafafi, Z. H., & Snow, A. W. (1989). Off-resonant third-order optical nonlinearities of metal-substituted phthalocyanines. *Applied Physics Letters, 55*(13), 1287–1288.

18. Bornacelli, J., Torres-Torres, C., Silva-Pereyra, H. G., Rodríguez-Fernández, L., Avalos-Borja, M., Cheang-Wong, J. C., & Oliver, A. (2017). Nanoscale influence on photoluminescence and third-order nonlinear susceptibility exhibited by ion-implanted Pt nanoparticles in silica. *Methods and Applications in Fluorescence, 5*(2), 025001.

19. Bornacelli, J., Torres-Torres, C., Can-Uc, B., Rangel-Rojo, R., Silva-Pereyra, H. G., Labrada-Delgado, G. J., & Oliver, A. (2019). Coupling effects and ultrafast third-order nonlinear optical behavior in ion-implanted silicon quantum dots and platinum nanoclusters. *Optical Materials, 97*, 109388.

20. Liu, C., Si, Y., Yang, G., & Pan, X. (2017). Chiral macrocyclic imine nickel (II) coordination complexes with diverse photophysical properties. *Dyes and Pigments, 140*, 70–78.

21. Trujillo, A., Fuentealba, M., Carrillo, D., Manzur, C., Ledoux-Rak, I., Hamon, J. R., & Saillard, J. Y. (2010). Synthesis, spectral, structural, second-order nonlinear optical properties and theoretical studies on new organometallic donor− acceptor substituted nickel (II) and Copper (II) unsymmetrical Schiff-base complexes. *Inorganic Chemistry, 49*(6), 2750–2764.

22. Tamer, Ö., Avcı, D., & Atalay, Y. (2015). Synthesis, X-ray structure, spectroscopic characterization and nonlinear optical properties of Nickel (II) complex with picolinate: A combined experimental and theoretical study. *Journal of Molecular Structure, 1098*, 12–20.

23. Cai, Z., Zhou, M., & Xu, J. (2011). Degenerate four-wave mixing determination of third-order optical nonlinearities of three mixed ligand nickel (II) complexes. *Journal of Molecular Structure, 1006*(1–3), 282–287.

24. Winter, C. S., Hill, C. A. S., & Underhill, A. E. (1991). Near resonance optical nonlinearities in nickel dithiolene complexes. *Applied Physics Letters, 58*(2), 107–109.

25. Chandra, M., & Das, P. K. (2009). Size dependence and dispersion behavior of the first hyperpolarizability of copper nanoparticles. *Chemical Physics Letters, 476*(1–3), 62–64.

26. Chandra, M., Indi, S. S., & Das, P. K. (2007). Depolarized hyper-Rayleigh scattering from copper nanoparticles. *The Journal of Physical Chemistry C, 111*(28), 10652–10656.

27. Uchida, K., Kaneko, S., Omi, S., Hata, C., Tanji, H., Asahara, Y., & Nakamura, A. (1994). Optical nonlinearities of a high concentration of small metal particles dispersed in glass: Copper and silver particles. *Journal of the Optical Society of America B, 11*(7), 1236–1243.

28. Cetin, A., Kibar, R., Hatipoglu, M., Karabulut, Y., & Can, N. (2010). Third-order optical nonlinearities of Cu and Tb nanoparticles in SrTiO3. *Physica B: Condensed Matter, 405*(9), 2323–2325.

29. Li, X., Li, S., Ren, H., Yang, J., & Tang, Y. (2017). Effect of alkali metal atoms doping on structural and nonlinear optical properties of the gold-germanium bimetallic clusters. *Nanomaterials, 7*(7), 184.

30. Dadge, J. W., Islam, M., Dharmadhikari, A. K., Mahamuni, S. R., & Aiyer, R. C. (2006). Hyper-Rayleigh scattering in electrochemically synthesized Ag–Au coupled clusters. *Journal of Physics: Condensed Matter, 18*(23), 5405.

31. Russier-Antoine, I., Bachelier, G., Sabloniere, V., Duboisset, J., Benichou, E., Jonin, C., & Brevet, P. F. (2008). Surface heterogeneity in Au-Ag nanoparticles probed by hyper-Rayleigh scattering. *Physical Review B, 78*(3), 035436.
32. Kirubha, E., & Palanisamy, P. K. (2014). Green synthesis, characterization of Au–Ag core–shell nanoparticles using gripe water and their applications in nonlinear optics and surface enhanced Raman studies. *Advances in Natural Sciences: Nanoscience and Nanotechnology, 5*(4), 045006.
33. Kim, M. J., Na, H. J., Lee, K. C., Yoo, E. A., & Lee, M. (2003). Preparation and characterization of Au–Ag and Au–Cu alloy nanoparticles in chloroform. *Journal of Materials Chemistry, 13*(7), 1789–1792.
34. Papagiannouli, I., Aloukos, P., Rioux, D., Meunier, M., & Couris, S. (2015). Effect of the composition on the nonlinear optical response of Au x Ag1–x nano-alloys. *The Journal of Physical Chemistry C, 119*(12), 6861–6872.
35. Ferreira, E., Kharisov, B., Vázquez, A., Méndez, E. A., Severiano-Carrillo, I., & Trejo-Durán, M. (2020). Tuning the nonlinear optical properties of Au@Ag bimetallic nanoparticles. *Journal of Molecular Liquids, 298*, 112057.
36. Zhang, Q. F., Liu, W. M., Xue, Z. Q., Wu, J. L., Wang, S. F., Wang, D. L., & Gong, Q. H. (2003). Ultrafast optical Kerr effect of Ag–BaO composite thin films. *Applied Physics Letters, 82*(6), 958–960.
37. Koirala, K. P., Garcia, H., Sandireddy, V. P., Kalyanaraman, R., & Duscher, G. (2020). Bimetallic Fe-Ag arrays with extraordinary nonlinear refraction and nonlinear Faraday rotation at telecommunication wavelength (1550 nm).
38. Haghighatzadeh, A. (2020). Enhanced third-order optical susceptibility in Ag-doped CeO2 nanostructures under pulsed Nd-YVO4 laser. *Optics & Laser Technology, 126*, 106114.
39. Zhong, J., Xiang, W., Chen, Z., Xie, C., Luo, L., & Liang, X. (2013). Microstructures and third-order optical nonlinearities of Cu_2In nanoparticles in glass matrix. *Journal of Alloys and Compounds, 572*, 137–144.
40. Zhang, Y., Jin, Y., He, M., Zhou, L., Xu, T., Yuan, R., & Liang, X. (2018). Optical properties of bimetallic Au-Cu nanocrystals embedded in glass. *Materials Research Bulletin, 98*, 94–102.
41. Huang, Y., Xiang, W., Yang, R., Yan, L., & Liang, X. (2017). Sol-gel derived glass nanocomposites embedded with phase-controlled Cu-Ni nanostructures and their optical non-linearities. *Materials Letters, 189*, 184–187.
42. Huang, Y., Xiang, W., Lin, S., Cao, R., Zhang, Y., Zhong, J., & Liang, X. (2017). The synthesis of bimetallic gold plus nickel nanoparticles dispersed in a glass host and behavior-enhanced optical nonlinearities. *Journal of Non-Crystalline Solids, 459*, 142–149.
43. Liu, M., Zhong, J., Ma, X., Huang, Y., & Xiang, W. (2019). Sodium borosilicate glass doped with Ni nanoparticles: Structure, formation mechanism and nonlinear optical properties. *Journal of Non-Crystalline Solids, 522*, 119560.
44. Zhong, J., & Xiang, W. (2017). Sol-gel synthesis and third-order nonlinear optical properties of Cu3. 8Ni nanoparticles doped glass. *Journal of Non-Crystalline Solids, 462*, 17–22.
45. Hernández-Acosta, M. A., Soto-Ruvalcaba, L., Martínez-González, C. L., Trejo-Valdez, M., & Torres-Torres, C. (2019). Optical phase-change in plasmonic nanoparticles by a two-wave mixing. *Physica Scripta, 94*(12), 125802.
46. Fernández-Valdés, D., Torres-Torres, C., Martínez-González, C. L., Trejo-Valdez, M., Hernández-Gómez, L. H., & Torres-Martínez, R. (2016). Gyroscopic behavior exhibited by the optical Kerr effect in bimetallic Au–Pt nanoparticles suspended in ethanol. *Journal of Nanoparticle Research, 18*(7), 204.
47. Pérez, J. L. J., Gutiérrez-Fuentes, R., Ramírez, J. F. S., Vidal, O. U. G., Téllez-Sánchez, D. E., Pacheco, Z. N. C., & García, J. A. F. (2013). Nonlinear coefficient determination of Au/Pd bimetallic nanoparticles using Z-scan. *Advances in Nanoparticles, 2*(03), 223–228.
48. Brodyn, M. S., Volkov, V. I., Rudenko, V. I., Liakhovetskyi, V. R., & Borshch, A. O. (2017). Large third-order optical nonlinearity in PdO thin films. *Journal of Nonlinear Optical Physics & Materials, 26*(03), 1750037.

Chapter 5
Study on Second- and Third-Order Nonlinear Optical Properties in Semiconductor Nanoparticles and Quantum Dots

5.1 Semiconductor Nanostructures: Properties and Classification

Semiconductor nanomaterials also attract characteristic attention for application design and development of all-optical functions. Semiconductor nanomaterials possess good linear absorption, enhanced optical emission and photoluminescence, as well as excellent nonlinear optical response due to quantum confinement effect. In order to control the physicochemical properties exhibited by semiconductor nanomaterials, it is convenient to modify parameters such as the size, shape, and surface characteristics of the nanoparticles. In addition, synthesis, characterization, and measurement of the properties allow identifying the suitable elements for the preparation of potential hybrid semiconductors for optoelectronic and photonic applications. One of the remarkable properties of semiconductor nanomaterials is based on quantum confinement.

The spatial quantum confinement effect described in Fig. 5.1 results in significant modification to the optical properties of semiconductor nanomaterials. Similarly, there is an important relationship between the surface of the nanomaterial and its volume, which influences the optical and surface properties of the material; quantum phenomena are widely manifested in semiconductors, due to the fact that the energy levels and the separation between said energy levels increase as the size of the particle decreases.

The structures of the electronic bands of metals, semiconductors, and dielectric materials show significant differences. The valence band and conduction band of metals are filled with electrons that can move freely through the conduction band so that there is no separation between the electronic bands. On the other hand, in semiconductors there is a bandgap between both bands.

The effect of light emission that occurs in semiconductor materials is due to their geometry. Semiconductors can be represented as a crystalline solid that has a system

© The Author(s), under exclusive license to Springer Nature Switzerland AG 2022

C. Torres-Torres, G. García-Beltrán, *Optical Nonlinearities in Nanostructured Systems*, Springer Tracts in Modern Physics 287, https://doi.org/10.1007/978-3-031-10824-2_5

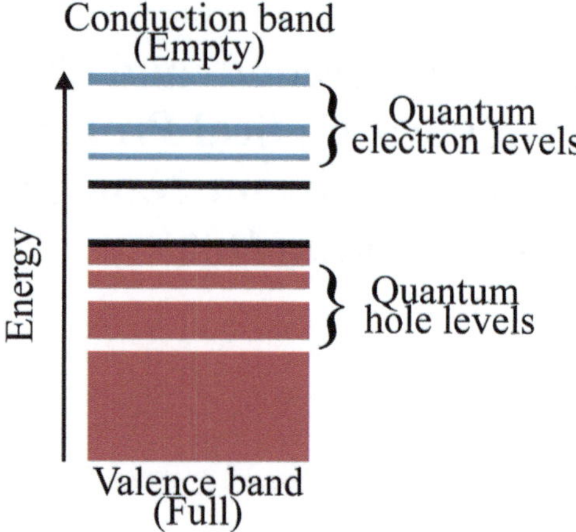

Fig. 5.1 Diagram of the quantum confinement effect on electron density states

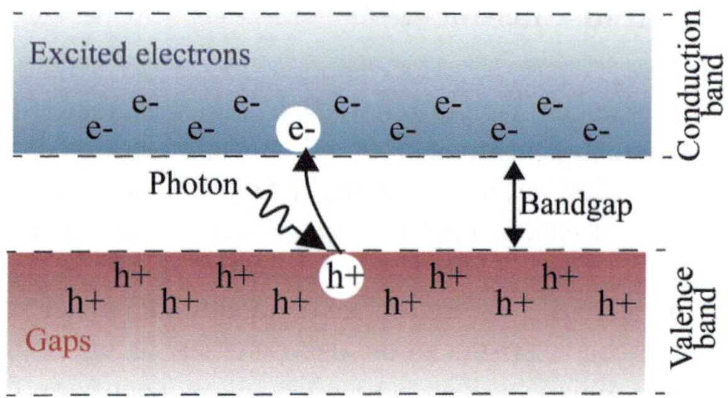

Fig. 5.2 Diagram of semiconductor energy band

of energy bands known as the valence band and conduction band. Both bands are separated by a gap known as bandgap, which represents forbidden energies for electrons. If an electromagnetic field is applied on the system, the electrons contained in the valence band reach an excited state with the necessary energy to go toward the conduction band. This means that a positive charge gap will be left in the valence band. On the other hand, the electrons excited in the conduction band will seek to return to the valence band to regain equilibrium and occupy one of the available holes. The process explained above is known as electron–hole pair recombination and is described in Fig. 5.2. Moreover, it is necessary for the electrons to release the energy acquired from the induced electromagnetic wave. So, the emitted wavelength is directly related to the energy of the bandgap in the semiconductor material.

On the other hand, the dielectric properties of semiconductors associated with the optical properties result in a dielectric constant that depends on the frequency; dielectric behavior is described as a sum of resonances. Each of the resonances occurs at a specific frequency; correspondingly, the spatial dispersion effect is an effect resulting from the dependence of the wave vector of the resonant frequencies on the optical properties of semiconductor nanostructures. Thus, for the design of applications in areas that include energy conversion, sensing, electronics, photonics, and biomedicine, the development of hybrid nanomaterials with improved optical and electronic properties is necessary.

The chemical structure of semiconductor materials is identified because the lower level of the valence band is almost full of electrons. On the contrary, the upper level of the conduction band remains almost empty. Because of this, it is possible for electrons to jump between the bands, passing through the energy band. Thus, semiconductor materials exhibit confining effects. These effects occur because the energy levels between the conduction and valence band balance out and can be treated as continuous phases. One of the factors responsible for the quantum confinement of semiconductor materials is the density due to the size of the particles.

Thus, due to the size distribution of the particles, there is an increase in quantum confinement since semiconductor crystals have large bandgaps. Quantum confinement develops geometric structures close to the atom called quantum dots as described in Fig. 5.3. In quantum dots, the typical electronic transitions between the conduction band and valence band are exhibited; however, the size of the bandgap is adjusted through the size of the quantum dot. Then, the frequency of emission of the point is dependent on the interval of the band. Thus, quantum dots with small geometries produce light with short wavelengths. In contrast, quantum dots with large geometries produce light with long wavelengths. The described mechanism is interpreted through the visible electromagnetic spectrum.

The confinement of electrons in small metallic materials is one of the main factors in changing the optical response. The size of the materials has an important influence on the nonlinear optical response. So, if the size of the material is less than the wavelength of the applied radiation, it results in the collective oscillation of electrons that share the same homogeneous electromagnetic field. This excitation can be called resonant when the applied wave frequency coincides with the natural frequency of the movement of the electron gas in relation to the ionic nucleus. The phenomenon of resonance associated with the derived metallic nanostructured materials has important optical characteristics such as spectral scattering and absorption, in addition to strong near-field electromagnetic enhancements.

5.2   Silicon Second- and Third-Order Optical Nonlinearities

Other interesting materials due to their nonlinear third-order optical properties are the implanted ionic nanocomposites. Furthermore, as previously stated, quantum dots are a clear example in applications of optical processes related to third-order nonlinear susceptibility. Thus, implanted compounds of silicon quantum dot ions are also of interest in the field of optical nonlinearities. In the study of silicon quantum dot compounds, nonlinearities have been exhibited in the nanosecond and femtosecond regime. Ultrafast nonlinearities in the fs regime are associated with the optical Kerr effect, while a thermal effect is identified as the main mechanism responsible for the nonlinear optical refraction induced by nanosecond pulses.

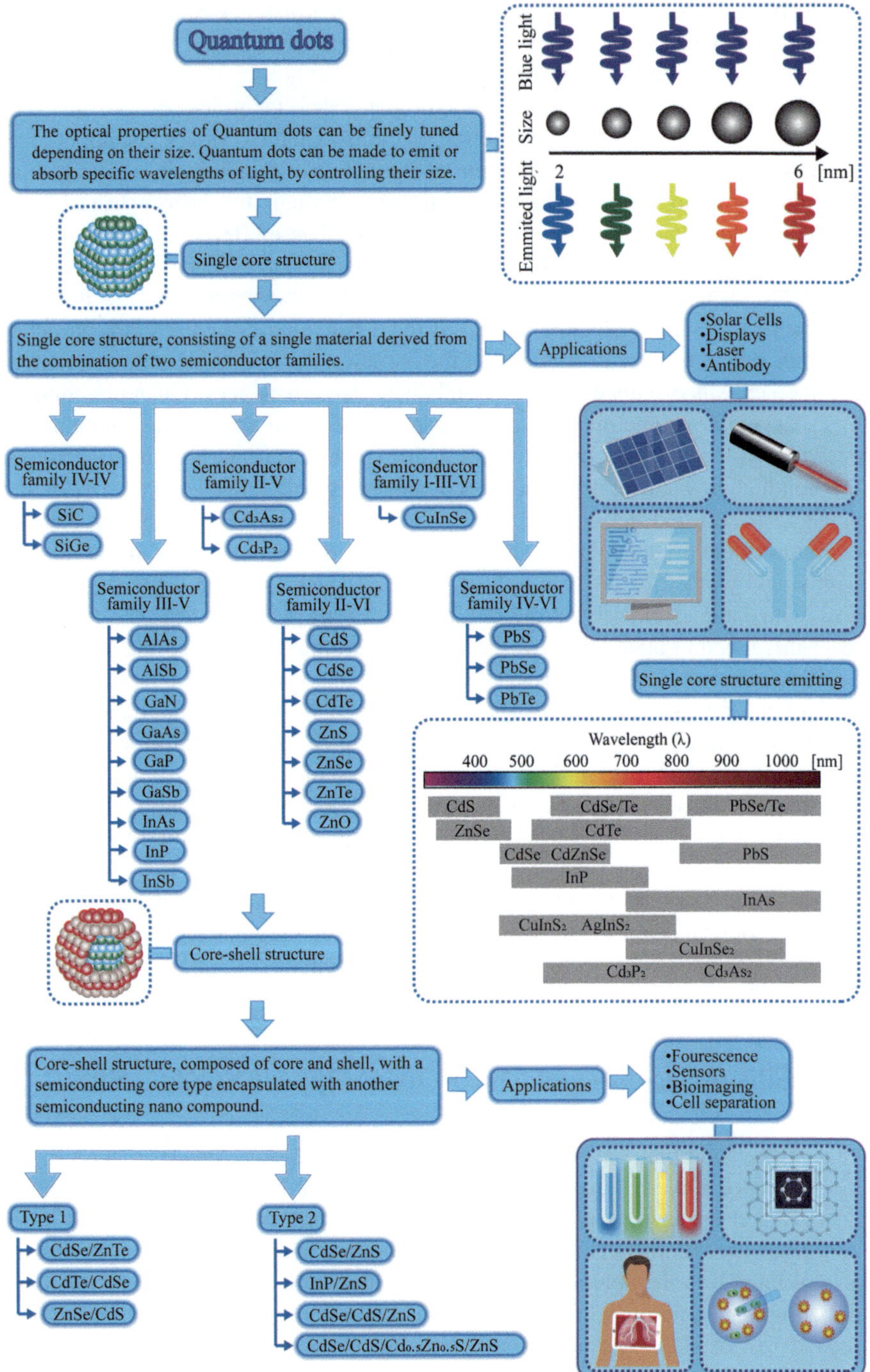

Fig. 5.3 Graphical synthesis of quantum dots semiconductor properties and applications

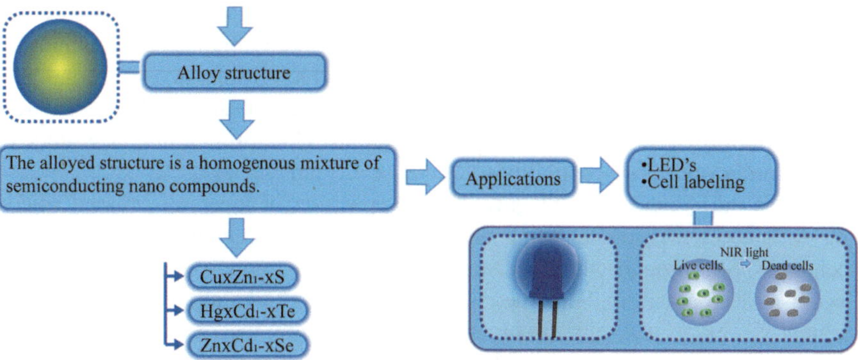

Fig. 5.3 (continued)

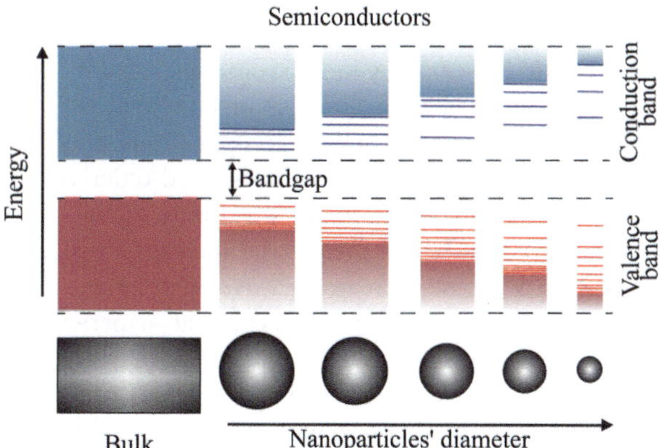

Fig. 5.4 Diagram of the energy change of the bandgap for semiconductors of different diameters

Quantum confinement effects occur in semiconductor materials in nanometric sizes. The minimum diameter required for quantum confinement effects is different for each material. Furthermore, through the quantum confinement of the semiconductors it is possible to increase the efficiency of light emission in the semiconductors, in addition to the fact that it is possible to modulate the emission length from the ultraviolet to the infrared. The modulation of the emission length of semiconductors depends on the diameter of the semiconductor nanocrystal. Thus, reducing the diameter of the semiconductor nanocrystal induces an increase in the bandgap of the semiconductor as shown in Fig. 5.4 such that a wider bandgap tends to move in the direction of the blue-violet colors.

Optical susceptibilities associated with second-order optical effects are necessary for single-phase modulation processes, in addition to second harmonic generation and frequency sum and difference generation. It is known that, due to its symmetry, crystalline silicon does not present a second-order nonlinear optical susceptibility response, $\chi^{(2)}$. Thus, symmetry breaking is necessary to observe second-order non-linearities in silicon waveguides. The breaking of the symmetry has been carried out through the application of current fields, such that a second-order optical susceptibility $\chi^{(2)}$ is induced. Thus, the obtained second-order optical susceptibility is exploited for the perturbation of the permittivity, which results in a phase modulation. The spatial distribution of the second-order optical susceptibility $\chi^{(2)}$ was measured through the second harmonic generation. This resulted in a magnitude of second-order optical susceptibility $\chi^{(2)} = 41 \pm 1.5$ pm V^{-1} [1].

On the other hand, the influence of the deposition of layers of stressed silicon nitride has been investigated using measurements of the second harmonic generation technique. So, through inhomogeneous deformation in silicon, the contribution of second-order optical nonlinearities is demonstrated. The breaking of symmetry in geometry is associated with the stress gradient in silicon, whereas the charge on the silicon nitride layers is responsible for a pronounced contribution of the second harmonic induced by the electric field. It has been observed that there is an improvement in second-order nonlinearity due to a constructive overlap of stress-induced effects and related to the deposition of silicon nitride layers [2].

In addition, a distinction has been made between the dipole contributions associated with non-centrosymmetric structures and the quadrupole contribution, which is also associated with centrosymmetric materials. Also, there is a wavelength dependence of the second-order optical susceptibility $\chi^{(2)}$ induced in deformed silicon waveguides such that the second-order optical susceptibility dependence $\chi^{(2)}$ has a magnitude of 340 pm V^{-1} for its silicon-covered waveguides. In addition, it has been shown that there is an increase in second-order optical susceptibility associated with the increase in wavelength. So, at larger wavelengths, the field is stronger in the waveguide facets that experience the strong inhomogeneous deformation [3].

Additionally, a large nonlinear second-order optical response $\chi^{(2)}$ has been observed in thin silicon carbide films on sapphire and fused silicone substrates such that the silicone samples deposited on the sapphire substrate have exhibited a magnitude of 10 pm V^{-1} [4].

High second-order nonlinearities have been measured in silica glass samples. The samples have been prepared through a sol–gel process with pressure variations. Thus, the samples have demonstrated a magnitude of second-order optical susceptibility $\chi^{(2)}$ in the range of 0.29–1.07 pm V^{-1}. Thus, through this method it has been shown that it is possible to induce second-order nonlinearities by thermal polishing on silica prepared by the sol–gel method [5].

Then, from these evaluations, it has been found that the nonlinear third-order optical response associated with ionic nanocomposites implanted with quantum dots of silicon exhibits a value of $\chi^{(3)} = 8.68 \times 10^{-9}$ esu at wavelength of 825 nm and 80 fs. Whereas at a wavelength of 532 nm and pulses of 1 ns, a value of $\chi^{(3)} = 5.3 \times 10^{-10}$ esu was exhibited. On the other hand, results have also been studied for ionic nanocomposites implanted with quantum points of silicon and nanoparticles of Au. In these compounds, a value of $\chi^{(3)} = 2.65 \times 10^{-9}$ esu has been demonstrated at a wavelength of 825 nm and 80 fs. Whereas at a wavelength of 532 nm and pulses of 1 ns, a value of $\chi^{(3)} = 3.5 \times 10^{-9}$ esu was exhibited [6]. As can be deduced, there are important differences in the nonlinearities of the two different samples in which the improvements are presented in the response studied in the ns regime, while a decrease in the resulting addition is observed for femtosecond pulses.

The third-order nonlinear susceptibility of semiconductor materials describes the independent contributions of valence electrons and conduction electrons. A comparison of third-order nonlinear optical responses between some semiconductor nanomaterials and Si shows that Si has a greater susceptibility to polarized radiation. The nonlinear third-order susceptibility of Si exhibits a value of $\chi^{(3)} = 0.08 \times 10^{-10}$ esu [7].

Through the coupling technique of two optical waves, it is also possible to measure the nonlinear third-order dynamic susceptibility of Si nanocomposites under resonance conditions. The electronic mechanism is responsible for generating nonlinearities in Si composite films. Moreover, the third-order nonlinear response has depended on factors such as the Si concentration in the sample films. Time-dependent measurements of the nonlinear response of the compound have shown a value of $\chi^{(3)} = 4 \times 10^{-7}$ esu in an ultrafast response regimen [8]. In this way, Si nanocomposite films are positioned on the list of interesting materials for optical applications since their third-order susceptibility value exhibits up to eight orders of magnitude greater than those of other compounds such as polyacetylene.

Semiconductor materials exhibit promising third-order optical nonlinear magnitudes due to their chemical structure that is due to the semiconductor materials having quantum confinement effects. These effects are because the energy levels between the conduction band and valence band balance out and can be treated as continuous phases.

5.3 Germanium Second- and Third-Order Optical Nonlinearities

In the case of germanium, quantum confinement phenomena present a larger aspect than other semiconductor nanomaterials such as silicon. Respectively, it is possible to appreciate quantum confinement effects in germanium nanoparticles smaller at 24 nm. Also, the nonlinear third-order response of semiconductors is directly related to the quantum confinement present.

The design of hybrid materials has been important due to the possibility of controlling linear and nonlinear optical interactions through them. From this, a silver/germanium compound has been studied. The interaction between the atoms of both materials exhibits strong resonances of surface plasmons located in grains of silver encapsulated with germanium. These result in the satisfactory second harmonic generation and second-order nonlinearities with up to two orders of magnitude higher compared to the thin films of silver without germanium atoms. The good nonlinear optical response derived from these compounds is related to the encapsulation process of silver atoms, such that the effective surface area necessary for a second-order nonlinear response increase [9].

On the other hand, the nonlinear coefficients of the second order exhibited due to the transitions between subbands in wells of multiple germanium/silicon quantum have been studied such that the magnitude of the second-order susceptibility has exhibited a significant value of 20,000 pm V^{-1} [10].

In addition, the response of the second-order nonlinearities of germanium quantum dots has also been studied. Thus, due to a symmetry break due to surface termination, the second-order optical susceptibility $\chi^{(2)}$ increases up to 299.1 pm V^{-1} [11].

Germanium has been shown to be a sensitive optical material, and its use can lead to optical enhancements in nanoantennas. Furthermore, germanium has a high optical susceptibility exposed to optical wave coupling phenomena. The nonlinear response of the germanium nanodisks is coupled to a single third-order resonant mode. However, due to the geometry of the disk there is a diminished near-field spatial overlap of the mixed wavelengths within the disk. Thus, the enhancement of the induced field within the germanium disk due to third-order nonlinear susceptibility produces values of $\chi^{(3)} = 2 \times 10^{-8}$ esu [12]. It is then observed that germanium nanodisks are useful for the optimization of third-order optical processes in totally dielectric nanostructures.

On the other hand, the third-order nonlinear optical properties of the Ge nanoparticles measured through the multi-wave mixing technique at a wavelength of 532 nm and ps pulses were estimated at $\chi^{(3)} = 1.1 \times 10^{-11}$ esu [13].

Besides, the study of nanoparticles suspended in various organic solvents is of vital importance due to the third-order susceptibilities that can add to the response of nanoparticles. Thus, optical responses of germanium nanoparticle suspensions have been studied. The nonlinear optical response occurs due to the quantum confinement generated from the change in the bandgap of the suspended germanium nanoparticles compared to bulk semiconductors. Therefore, in the suspensions there is a self-diffraction due to the optical Kerr effect induced by the laser beam, which triggers the response of the nonlinear refraction of the samples. In contrast, the nonlinear absorption exhibited is totally dependent on the absorption of two photons. Then, the third-order nonlinear susceptibility of the germanium suspensions is defined as $\chi^{(3)} = 5.3 \times 10^{-8}$ esu [14].

5.4   Zinc Oxide Second- and Third-Order Optical Nonlinearities

Zinc oxide is a semiconductor material that has generated interest due to its physical properties that favor the use of the material in the development of components for optical and electronic applications. These optical and electrical properties make zinc oxide an optoelectronic material with promising qualities for application development.

In this way, second- and higher-order optical nonlinearities have been studied in aluminum-doped zinc oxide nanostructures. In addition, spectral locations of the second harmonic generations have been exhibited at 775 nm, with a relative intensity of 1.15×10^{-7} [15].

On the other hand, nonlinear second-order optical studies of ZnO nanostructures have been carried out through the second harmonic generation using pulses of 40 fs at a wavelength of 800 nm. Through this, a nonlinear refractive index with a magnitude of 4×10^{-11} cm^2 W^{-1} and a nonlinear absorption coefficient of 8×10^{-7} cm W^{-1} have been exhibited. It is also observed that the generation of high harmonics is possible through the propagation of femtosecond pulses through the ZnO nanoparticles [16].

Additionally, zinc oxide nanostructures and their second-order nonlinear optical properties have been studied. Furthermore, the value associated with the first optical hyperpolarizability has been studied such that said the parameter exhibits values from 830 au to 4,954,509 au [17].

Additionally, the hyper-Rayleigh scattering technique has been used to measure the first-order optical hyperpolarizability of ZnS nanocrystals with a diameter of 2.5 nm. The results show a value of the first optical hyperpolarizability of 2.34×10^{-27} esu. It is also observed that there is an increase of at least two orders of magnitude in the value of the first hyperpolarizability of ZnS nanostructures compared to bulk ZnS. Also, optical nonlinearities are associated with volume and surface contributions due to the special surface structure of nanocrystals. On the other hand, other possible contributions such as the surface static electric field, solvent field, bulk contribution, and surface contribution are analyzed [17].

Many of the engineered applications use zinc oxide in thin film nanocomposites. That is why studies have been carried out to determine the responses of nonlinearities of monolayer and multilayer films of different thicknesses of zinc oxide. The results derived from third-order nonlinear optical response measurements at a wavelength of 532 nm for zinc oxide films exhibit $\chi^{(3)} = 4.4 \times 10^{-9}$ esu for 250 nm monolayer and $\chi^{(3)} = 1.9 \times 10^{-9}$ esu for 500 nm monolayer. Whereas, in the case of zinc oxide bilayers, a value of $\chi^{(3)} = 3.7 \times 10^{-9}$ esu is exhibited with a thickness of 500 nm and $\chi^{(3)} = 1.8 \times 10^{-9}$ in bilayers with a thickness of 1000 nm [18]. Regarding the measured results, it is considered that the morphology of the zinc oxide films causes the modification of the third-order nonlinear optical response through the coupling of optical waves. Furthermore, the thickness of the films causes improvements in the nonlinear responses of the material.

As observed, the nonlinear optical response can depend on the morphology and also on the size of the nanoparticles. The zinc oxide nanofluids describe optical responses that have been due to the increase in the size of the suspended nanoparticles. Furthermore, zinc oxide nanofluids exhibit quantum confinement mechanisms.

The results of the nonlinear optical response of zinc oxide nanofluids against the size of the nanoparticles in the range of 6–18 nm show a minimum value of $\chi^{(3)} = 1.3 \times 10^{-10}$ esu for nanoparticles with a size of 6 nm. On the other hand, the maximum value studied for 18 nm nanoparticles is $\chi^{(3)} = 9.4 \times 10^{-10}$ esu [19].

On the other hand, the temperature of the zinc oxide nanofluid preparation process influences the structure of the zinc oxide nanoparticles. The crystallinity and diameter of zinc oxide nanoparticles increase with reaction temperature.

Then, it is defined that temperature is an influencing factor in the nonlinear optical properties of the suspended zinc oxide nanoparticles. Besides, the energy bandgap of the nanoparticles increases with the decrease in the size of the nanoparticles due to the phenomenon of quantum confinement generated. Then, the third-order nonlinear optical response of the zinc oxide nanofluids reveals values of $\chi^{(3)} = 0.93 \times 10^{-10}$ esu for 9.6 nm nanoparticles, $\chi^{(3)} = 1.53 \times 10^{-10}$ esu for 11.2 nm nanoparticles, $\chi^{(3)} = 1.78 \times 10^{-10}$ esu for 12.4 nm nanoparticles, and $\chi^{(3)} = 1.83 \times 10^{-10}$ esu for 17.7 nm nanoparticles [20].

Optical wave coupling influences the identification of two-photon absorption processes and the optical Kerr effect. In addition, through this method it is possible to identify nonlinear optical signals propagated through the different regions of the sample material.

The characteristics of the third-order nonlinear optical response must be analyzed considering the modifications due to the high irradiance on the samples. Therefore, it is necessary to determine the amount of energy required to damage the samples, and that is why it is important to measure the ablation threshold in the materials. The third-order nonlinear optical susceptibility response exhibited by zinc oxide nanoparticles studied below the ablation threshold of the samples presents a value of $\chi^{(3)} = 2.55 \times 10^{-9}$ esu [21]. The physical characteristics and nonlinear optical effects of zinc oxide appear to be suitable for use in dynamic signal and data processing applications.

5.5   Titanium Dioxide Second- and Third-Order Optical Nonlinearities

Titanium oxide nanocomposites are interesting in the field of optoelectronic applications. Specifically, hybrid polymer nanocomposites with semiconductors have been of interest due to their high optical response. In addition, third-order nonlinear optical responses in semiconductor and hybrid crystalline nanoparticles have received attention.

Nanomaterials based on titanium dioxide have attractive optical properties for the design of applications related to photocatalysis. Thus, the optical properties related to the second-order nonlinear response of titanium dioxide nanostructures due to the application of an external electric field have been studied. Also, the external electric field has a great influence on nonlinear optical properties. The bandgap of titanium dioxide nanostructures begins to decrease with increasing electric field; through this, it is possible to measure the optical coefficient related to the first second-order optical hyperpolarizability. Thus, titanium dioxide samples exposed to an increase in electric field were studied and the optical hyperpolarizability exhibited magnitudes in the range of 0.31–1316.331×10^{-30} esu, with an electric field intensity in the range of 0–0.035 au [22].

Titanium dioxide has centrosymmetric geometry; however, it has a great response of the first hyperpolarizability in solution with nanoparticles of 10 nm in diameter. Measurements in representative samples were made using the hyper-Rayleigh scattering technique and considering a constant concentration. Then, the values of the first hyperpolarizability were estimated with a magnitude of 4.3×10^{-26} esu [23].

Likewise, in another study, the second-order nonlinearity of titanium dioxide nanoparticles has been studied through hyper-Rayleigh dispersion spectroscopy, in which the addition of a sensitizing dye is analyzed. However, the spectral tint extinguishes the second harmonic generation [24].

On the other hand, the titanium dioxide nanoparticles synthesized through the sol–gel method have shown good nonlinear second-order behavior. The above has been demonstrated through a high value of the nonlinear coefficient of the first hyperpolarizability, which has exhibited a value of 5.3×10^{-26} esu. Also, the results show that intensities of the second harmonic generation can be increased by increasing the voltage or pumping concentration of the nanoparticle [25].

The third-order nonlinear optical susceptibility of crystalline titanium dioxide nanoparticles dispersed in silicon films has presented a magnitude of $\chi^{(3)} = 4.1 \times 10^{-11}$ esu. The above measurement was with wavelength parameters of 1060 nm and pulses in the ns regime.

On the other hand, with measurement parameters of 780 nm and pulses of 250 fs, ultrafast optical nonlinearities have been observed in thin films nanocomposite with titanium oxide. The measured optical nonlinearity has a magnitude of $\chi^{(3)} = 1.7 \times 10^{-9}$ esu and a recovery time of 1.5 ps [26].

Titanium dioxide is used in advanced oxidation processes and is in the form of a nontoxic insoluble white solid. Titanium dioxide can be found in various mineral forms. The form known as rutile is found in a dark form. This mineral contains 98% titanium, while it does not present impurities and is white in color. On the other hand, anatase is not as hard and is less dense than rutile, and its luster is more adamantine. Anatase is a secondary mineral obtained from other titanium carrier minerals.

Anatase and rutile titanium dioxide films have been of interest because it is important to define whether any of the third-order optical nonlinearities are directly related to the crystalline structures of titanium dioxide. In fact, the nonlinear optical properties in titanium dioxide films have been studied by third harmonic generation and the values of third-order optical susceptibility were reported to be $\chi^{(3)} = 4.0 \times 10^{-12}$ esu for rutile structure films. On the other hand, third-order optical susceptibility for anatase films exhibited a value of $\chi^{(3)} = 2.4 \times 10^{-12}$ esu [27].

In addition, it is important to determine the values of the optical bandgap of the materials that provide quantum confinement. Therefore, the third-order nonlinear susceptibilities of the films of crystalline structures of anatase and rutile titanium dioxide have a wavelength of 800 nm with a duration of 50 fs. The results of the measurements express that the real and imaginary parts of the third-order nonlinear optical susceptibility $\chi^{(3)}$ of anatase films are -7.1×10^{-11} esu and -4.42×10^{-12} esu, respectively. Moreover, for the films of rutile crystalline structures, it is determined that the real and imaginary parts of third-order nonlinear optical susceptibility $\chi^{(3)}$ are -3.6×10^{-11} esu and -2.2×10^{-11} esu, respectively [28].

Gold and silver nanoparticles embedded in dielectric matrices are related to localized surface plasmon resonance absorption and are of interest in the field of nonlinear optics. Thus, such hybrid materials have been widely studied as nonlinear third-order optical materials of the resonant type. Furthermore, the third-order optical nonlinearities of composite materials are influenced by the type and size of embedded metal, dielectric constant, and thermal conductivity of the dielectric matrices of the semiconductor material. Then, the nonlinear third-order optical susceptibility value of titanium dioxide films with embedded Au nanoparticles exhibited a value of $\chi^{(3)} = 3.6 \times 10^{-7}$ esu [29].

5.6 Cadmium Telluride Second- and Third-Order Optical Nonlinearities

Materials based on semiconductor nanoparticles are very attractive due to their responsive physical properties, such as their geometry, emission tuning, photostability, and photoluminescence. These materials have photophysical properties superior to other organic nanomaterials. However, temperature is an important factor in the synthesis of some semiconductor nanomaterials. The importance of studying the physical and chemical properties of semiconductor nanomaterials is due to the fact that, by controlling the chemical structure and dimensions of nanomaterials, it is possible to adapt the linear and nonlinear optical properties of nanomaterials.

On the other hand, the magnitudes of the second-order nonlinear optical susceptibilities of cadmium telluride have been analyzed through pulses with a wavelength of 28.0 μm from which second-order nonlinearity results with magnitude of 5.9×10^{-11} m/V [30] have been obtained.

On the other hand, measurements of the second-order nonlinear optical response of semiconductor ternary compounds based on $Cd_{0.8}Zn_{0.2}Te$ and $Cd_{0.78}Mn_{0.22}Te$ have been made through the generation of second harmonic with phase mismatch spectrally resolved at a wavelength of 151.5 μm. Thus, the samples of the compounds have exhibited values of second-order optical nonlinearities in the range of 78–71 pm V^{-1} [31].

Then, if in the process of the synthesis of the materials a difference is controlled to modify the chemical structure of the samples, it is possible to obtain differences in the optical response of the materials synthetized. In the specific case of cadmium telluride nanoparticles, it is possible to use different protection agents in the synthesis process for the formation of nanocrystals of the material. Third-order nonlinear optical responses have been studied at a wavelength of 532 nm. Results have been measured for mercaptoacetic acid, mercaptopropionic acid, and 2-mercaptoethanol as protection agents as a stabilizer in the process of synthesis of cadmium telluride nanocrystals. It is well known that one of the most important techniques for the characterization of third-order optical nonlinearities of materials is the coupling of optical waves. The third-order nonlinear optical response of cadmium telluride nanoparticles synthesized with mercaptoacetic acid shows a value of $\chi^{(3)} = 26.64 \times 10^{-6}$ esu. On the other hand, for cadmium telluride synthesized with mercaptopropionic acid it shows a value of $\chi^{(3)} = 27.75 \times 10^{-6}$ esu. Finally, for the synthesis of cadmium telluride nanoparticles with the agent 2-mercaptoethanol, $\chi^{(3)} = 32.51 \times 10^{-6}$ esu was exhibited [32]. The values of the third-order optical susceptibility measurements of the cadmium telluride nanoparticles are related to the movement of the π electron cloud from the donor to the receptor that allows the molecule to polarize. Furthermore, measurements of the optical parameters in cadmium telluride nanoparticles have exhibited favorable results related to absorption spectra, photoluminescence, and quantum effects. On the other hand, the effect resulting from the coupling of waves impinged on a colloidal cadmium telluride sample at a nonresonant wavelength revealed the effective third-order nonlinear susceptibility.

The third-order nonlinear susceptibility of cadmium telluride nanoparticles suspended in toluene establishes a value of between $\chi^{(3)} = 2.08 \times 10^{-14}$ esu, a $\chi^{(3)} = 4.30 \times 10^{-15}$ esu, at different concentrations [33]. The hyperpolarizabilities indicate a contribution of an electronic polarization process to the third-order optical nonlinearity of the cadmium telluride nanoparticles in suspension. Therefore, the third-order optical nonlinearity response is one of electronic contribution due to the anharmonic oscillation of the bound electrons of cadmium telluride in suspension. The third-order nonlinear optical response of glasses doped with cadmium telluride nanoparticles has also been measured through the mixing of optical waves at a wavelength of 580 nm. The magnitude of third-order nonlinear susceptibility of the doped glass samples was estimated at a value of $\chi^{(3)} = 4 \times 10^{-7}$ esu [34]. A high absorption value is also recorded for cadmium telluride-doped glasses associated with the size of the nanocrystals of the material. This phenomenon is due to the quantum confinement effect.

Furthermore, the response of the nonlinearities of cadmium telluride nanoparticles has been observed through a mixing of optical waves and ps pulses at a wavelength of 1064 nm. The results of the measurements are attributed to the phenomena of two-photon absorption and the optical Kerr effect. The magnitude of third-order nonlinear optical susceptibility of the cadmium telluride nanoparticles exhibits a value of $\chi^{(3)} = 7.2 \times 10^{-10}$ esu [35].

References

1. Timurdogan, E., Poulton, C. V., Byrd, M. J., & Watts, M. R. (2017). Electric field-induced second-order nonlinear optical effects in silicon waveguides. *Nature Photonics, 11*(3), 200–206.
2. Schriever, C., Bianco, F., Cazzanelli, M., Ghulinyan, M., Eisenschmidt, C., de Boor, J., & Schilling, J. (2015). Second-order optical nonlinearity in silicon waveguides: Inhomogeneous stress and interfaces. *Advanced Optical Materials, 3*(1), 129–136.
3. Cazzanelli, M., & Schilling, J. (2016). Second order optical nonlinearity in silicon by symmetry breaking. *Applied Physics Reviews, 3*(1), 011104.

4. Lundquist, P. M., Ong, H. C., Lin, W. P., Chang, R. P. H., Ketterson, J. B., & Wong, G. K. (1995). Large second-order optical nonlinearities in pulsed laser ablated silicon carbide thin films. *Applied Physics Letters, 67*(20), 2919–2921.

5. Pruneri, V., Bonfrate, G., Kazansky, P. G., Takebe, H., Morinaga, K., Kohno, M., & Takeuchi, T. (1999). High second-order optical nonlinearities in thermally poled sol-gel silica. *Applied Physics Letters, 74*(18), 2578–2580.6.

6. Torres-Torres, C., López-Suárez, A., Can-Uc, B., Rangel-Rojo, R., Tamayo-Rivera, L., & Oliver, A. (2015). Collective optical Kerr effect exhibited by an integrated configuration of silicon quantum dots and gold nanoparticles embedded in ion-implanted silica. *Nanotechnology, 26*(29), 295701.

7. Wynne, J. J. (1969). Optical third-order mixing in GaAs, Ge, Si, and InAs. *Physical Review, 178*(3), 1295.

8. Gang, F., Kasatani, K., Okamoto, H., & Takenaka, S. (2005). Large and fast resonant third-order optical nonlinearities of silicon 2, 3-naphthalocyanine bis (trihexylsilyloxide). *Chinese Physics Letters, 22*(7), 1687.

9. Stefaniuk, T., Olivier, N., Belardini, A., McPolin, C. P., Sibilia, C., Wronkowska, A. A., & Zayats, A. V. (2017). Self-assembled silver–germanium nanolayer metamaterial with the enhanced nonlinear response. *Advanced Optical Materials, 5*(22), 1700753.

10. Frigerio, J., Ballabio, A., Ortolani, M., & Virgilio, M. (2018). Modeling of second harmonic generation in hole-doped silicon-germanium quantum wells for mid-infrared sensing. *Optics Express, 26*(24), 31861–31872.

11. Islam, S., Shah, H., Shiri, D., Nekovei, R., & Verma, A. (2018). Ab-initio calculation of nonlinear optical susceptibilities in germanium quantum dots. In *2018 IEEE 13th Nanotechnology Materials and Devices Conference (NMDC)* (pp. 1–4).

12. Grinblat, G., Li, Y., Nielsen, M. P., Oulton, R. F., & Maier, S. A. (2017). Degenerate four-wave mixing in a multiresonant germanium nanodisk. *ACS Photonics, 4*(9), 2144–2149.

13. Sreeramulu, V., Saikiran, V., & Rao, D. N. (2015). Optical, structural and third-order nonlinear optical properties of Ge nanoparticles prepared by laser ablation in acetone. *Asian Journal of Physics, 24*(10), 1381–1389.

14. Ganeev, R. A., Ryasnyanskiy, A. I., & Usmanov, T. (2007). Optical and nonlinear optical characteristics of the Ge and GaAs nanoparticle suspensions prepared by laser ablation. *Optics Communications, 272*(1), 242–246.

15. Wu, J., Xie, Z. T., Fu, H. Y., & Li, Q. (2020). High-order harmonic generations in epsilon-near-zero aluminum-doped zinc oxide Nanopyramid Array. In *2020 12th International Conference on Advanced Infocomm Technology (ICAIT)* (pp. 5–9).

16. Rout, A., Boltaev, G. S., Ganeev, R. A., Rao, K. S., Fu, D., Rakhimov, R. Y., & Guo, C. (2019). Low-and high-order nonlinear optical studies of ZnO nanocrystals, nanoparticles, and nanorods. *The European Physical Journal D, 73*(11), 1–8.

17. Reddy, I. V., Baev, A., Prasad, P. N., & Agren, H. (2021). Pulsed response theory prediction of ZnO nanocluster polarizabilities: A benchmark study. *Chemical Physics Letters, 778*, 138746.

18. Torres-Torres, C., Castaneda, L., Trejo-Valdez, M., Maldonado, A., & Torres-Martinez, R. (2012). Participation of the third order optical nonlinearities in nanostructured silver doped zinc oxide thin solid films. *Journal of Nanomaterials, 2012*. https://doi.org/10.1155/2012/353061

19. Irimpan, L., Nampoori, V. P. N., Radhakrishnan, P., Krishnan, B., & Deepthy, A. (2008). Size-dependent enhancement of nonlinear optical properties in nanocolloids of ZnO. *Journal of Applied Physics, 103*(3), 033105.

20. Ramya, M., Nideep, T. K., Vijesh, K. R., Nampoori, V. P. N., & Kailasnath, M. (2018). Synthesis of stable ZnO nanocolloids with enhanced optical limiting properties via simple solution method. *Optical Materials, 81*, 30–36.

21. Ortíz-Trejo, F., Trejo-Valdez, M., Campos-López, J. P., Castro-Chacón, J. H., & Torres-Torres, C. (2020). Multipath data storage by third-order nonlinear optical properties in zinc oxide nanostructures. *Applied Sciences, 10*(16), 5688.

22. Zhang, X., Liu, Y., Ma, X., Jin, F., Abulimiti, B., & Xiang, M. (2020). Tuning the optical properties of (TiO2) 2 via the electric field. *Optik, 221*, 165395.
23. Fu, D. G., Zhang, Y., Wang, X., Liu, J. Z., & Lu, Z. H. (2001). Surface second order optical nonlinearity of titanium dioxide sized in nanometer range. *Chemistry Letters, 30*(4), 328–329.
24. Sahyun, M. R. V. (2002). Hyper-Rayleigh scattering (HRS) spectroscopy applied to nanoparticulate TiO2. *Spectrochimica Acta Part A: Molecular and Biomolecular Spectroscopy, 58*(14), 3149–3157.
25. Nafil, R. Q., & Majeed, M. S. (2020). Frequency doubling by nonlinearity of TiO2 nanomaterial. *Heliyon, 6*(3), e03649.
26. Elim, H. I., Ji, W., Yuwono, A. H., Xue, J. M., & Wang, J. (2003). Ultrafast optical nonlinearity in poly (methylmethacrylate)-TiO_2 nanocomposites. *Applied Physics Letters, 82*(16), 2691–2693.
27. Hashimoto, T., Yoko, T., & Sakka, S. (1994). Sol–gel preparation and third-order nonlinear optical properties of TiO2 thin films. *Bulletin of the Chemical Society of Japan, 67*(3), 653–660.
28. Long, H., Chen, A., Yang, G., Li, Y., & Lu, P. (2009). Third-order optical nonlinearities in anatase and rutile TiO2 thin films. *Thin Solid Films, 517*(19), 5601–5604.
29. Tanahashi, I., & Mito, A. (2011). Femtosecond optical nonlinearities of Au/TiO_2 thin films prepared by a sputtering method. *Journal of Materials Research, 26*(6), 763.
30. Sherman, G. H., & Coleman, P. D. (1973). Measurement of the second-harmonic-generation nonlinear susceptibilities of Se, CdTe, and InSb at 28.0 μm by comparison with Te. *Journal of Applied Physics, 44*(1), 238–241.
31. Zappettini, A., Pietralunga, S. M., Milani, A., Martinelli, M., & Andrzej, M. (2000). Measurements of second-order susceptibility a λ= 1.5 μm in CdTe-based ternary alloys efficient wavelenght conversion. *Journal of Applied Physics, 88*(8), 4913–4915.
32. Abd El-sadek, M. S., Nooralden, A. Y., Babu, S. M., & Palanisamy, P. K. (2011). Influence of different stabilizers on the optical and nonlinear optical properties of CdTe nanoparticles. *Optics Communications, 284*(12), 2900–2904.
33. Ma, S. M., Seo, J. T., Yang, Q., Battle, R., Creekmore, L., Lee, K., & Yu, W. (2007). The second hyperpolarizability of CdTe nanocrystals using polarization-resolved degenerate four-wave mixing. *Applied Surface Science, 253*(15), 6612–6615.
34. Ohtsuka, S., Koyama, T., Tsunetomo, K., Nagata, H., & Tanaka, S. (1992). Nonlinear optical property of CdTe microcrystallites doped glasses fabricated by laser evaporation method. *Applied Physics Letters, 61*(25), 2953–2954.
35. Canto-Said, E. J., Hagan, D. J., Young, J., & Van Stryland, E. W. (1991). Degenerate four-wave mixing measurements of high order nonlinearities in semiconductors. *IEEE Journal of Quantum Electronics, 27*(10), 2274–2280.

Chapter 6
Study on Second- and Third-Order Nonlinear Optical Properties in Nanostructured Systems: Nanocrystals and Complex Geometries

6.1 Nanocrystals

Interest in nonlinear optical interactions in solids has grown exponentially because of the development of high-energy laser systems. Optical materials with crystalline structures normally behave as linear transmitters. This is described through a monochromatic beam of light that travels through the material and is transmitted proportionally to its intensity. The speed of the light beam can vary due to the composition of the material. On the other hand, modulation of the beam wavelength as illustrated in Fig. 6.1 is possible throughout its transmission through the material.

However, the frequency remains constant. On the contrary, if the light beam is of high intensity, the material exhibits nonlinearities. There are numerous structural types of crystalline compounds, allowing a potential to choose favorable structures for nonlinear optical applications. On the other hand, synthesis methods influence the optical quality of crystal structures.

An example of these processing routes corresponds to the structures synthesized through the high-temperature flow method. With such a synthesis method, it is also possible to achieve good properties such as a high ablation threshold, chemical stability, and transparency in the ultraviolet region of the electromagnetic spectrum. Materials with crystalline structures exhibiting nonlinearities attract attention due to their characteristics suitable for high-speed, optical and switching devices. On the other hand, it has also been shown that image amplification is possible through energy transfer derived from the mixing of optical waves with crystals since the intensity transmitted by the crystal is amplified 10 times as a result of the self-diffraction of the reference beam shifted through the dynamic medium. So, the interaction of optical waves can work in displaying patterns. Then, through the coupling of optical waves through crystals, applications can be designed in the field of holographic interferometry, sensors, and coherent wave amplification [1].

© The Author(s), under exclusive license to Springer Nature Switzerland AG 2022

C. Torres-Torres, G. García-Beltrán, *Optical Nonlinearities in Nanostructured Systems*, Springer Tracts in Modern Physics 287, https://doi.org/10.1007/978-3-031-10824-2_6

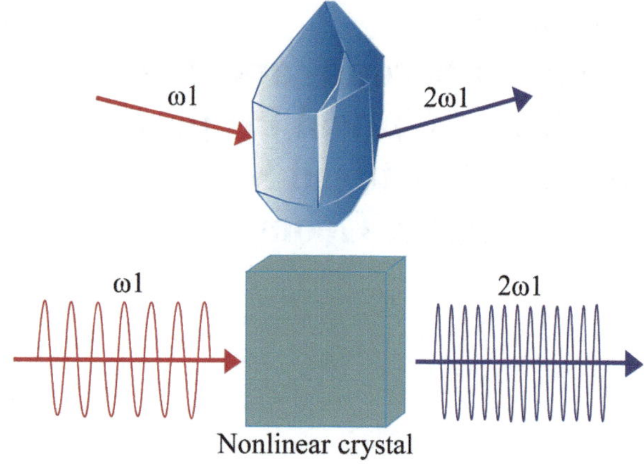

Fig. 6.1 Representation of modulation of the beam wavelength by the nonlinear crystal

Also, third-order nonlinear optical materials having the peculiarity of weak nonlinear optical absorption with strong nonlinear refraction attract important attention. These characteristics can be applied in the design of signal processing devices. There are a multitude of crystalline structures that use nonlinear optical phenomena for application design purposes, including light modulators, beam deflectors, frequency changers, second harmonic generators, and parametric oscillators and amplifiers.

Furthermore, crystal growth technology exhibits good enough nonlinear third-order optical susceptibilities for current photonic applications. However, they have the limitation of working only with monocrystalline materials. Besides, the optical switching time is relatively slow.

6.2 Lithium Crystals Second- and Third-Order Optical Nonlinearities

Optical applications originating due to nonlinear effects are due to electronic interactions within matter due to the irradiation of high-energy beams. These interactions have the quality of producing nonlinear optical effects and generating second-order or higher harmonics. Materials that involve knowledge of optical phenomena extend to semi-organic crystals due to their ability to combine the most prominent characteristics of organic and inorganic remains. Thus, studies based on the synthesis and growth of crystals as nonlinear optical materials are of importance due to their potential applications in the field of laser technology and quantum data storage.

The second-order optical nonlinearity is associated with the design of classical and quantum order photonic applications. In the same way, for the study of second-order nonlinear optical processes $\chi^{(2)}$ of thin film lithium niobate waveguides, a second harmonic field was identified with a wavelength of 775 nm. Through measurements, it is shown that the phase coincidence for the second-order optical nonlinearity in the direction of propagation is along the axis and lithium niobate can be performed simultaneously in the same waveguide manufactured from thin film lithium niobate with transverse electric fields and two transverse electric and transverse magnetic mode fields. Furthermore, the simultaneous measurement of both phase adaptation conditions is associated with the dependence of the polarization of the second harmonic generation. Thus, the advantage of making measurements through different phase coincidence conditions with a single photonic device is valuable in the design of multifunctional photonic devices [2].

Kerr microresonator frequency combs could stand out due to their wide range of applications. However, many of these applications require strong manipulation of the generated frequency comb signal using photonic elements with powerful second-order nonlinearity $\chi^{(2)}$. Due to this, the response of second-order optical susceptibility $\chi^{(2)}$ is studied through a monolithic compound of lithium niobate nanophotonic waveguides. Thus, optical nonlinearities have been observed from the second harmonic generation for a crystallized surface of lithium niobate glass 0.35 pm thick. The effective second-order nonlinear optical susceptibility of the lithium niobate film exhibited a magnitude of -5.25 pmV^{-1} [3].

Lithium niobate has large second-order nonlinearities $\chi^{(2)}$ with a magnitude of 3×10^{-11} m sV^{-1}, in addition to low optical losses. On the other hand, the second-order optical nonlinearity $\chi^{(2)}$ associated with the electro-optical effect can originate in order to manipulate the comb generated by an external electric field such that, due to the effective nonlinearity $\chi^{(2)}$ in a waveguide, the generation of frequency combs with a lower power threshold [4] is possible.

On the other hand, in another study the response of second-order optical nonlinearities was measured through the application of the hyper-Rayleigh scattering technique on lithium niobate particles such that the first-order optical hyperpolarizability exhibited a magnitude of 16.3×10^{-24} [5].

Lithium bis (2-methylactate) borate monohydrate is a semi-organic crystal that possesses attractive properties such as thermal stability, high linear optical transmittance, and also high third-order nonlinear efficiency. The third-order nonlinear optical susceptibility of lithium bis (2-methylactate) borate monohydrate shows a value of $\chi^{(3)} = 4.55 \times 10^{-5}$ esu [6]. The high value in third-order nonlinear optical susceptibility of lithium bis (2-methylactate) borate monohydrate is directly associated with the intermolecular interaction exhibited by the crystal structure in addition to the O–Li–O intermolecular interactions.

Through the generation of the optical Kerr effect on crystalline materials, it is possible to use the materials in photonic applications of nanometric sizes. Lithium niobate is a crystal that possesses nonlinear second- and third-order optical responses simultaneously, as well as low-energy losses. The lithium niobate crystal has a third-order nonlinear susceptibility response in a magnitude of $\chi^{(3)} = 1.15 \times 10^{-13}$ esu derived from the generation of the optical Kerr effect obtained through the spectral and temporal manipulation of the induced signal [4].

One of the main purposes of lithium niobate crystal studies is miniaturization for photonic component applications. Lithium niobate single crystals have also been studied in thin film form using the crystal ion-cutting technique. The nonlinear optical response of lithium niobate deposited on fused silica exhibits a value of $\chi^{(3)} = 2.5 \times 10^{-11}$ esu. The nonlinear optical response of lithium niobate is deposited on zinc oxide and fused silica with a magnitude of $\chi^{(3)} = 4.3 \times 10^{-11}$ esu. Therefore, the nonlinear optical response of lithium niobate deposited on zinc oxide and Z-cut quartz presents a value of $\chi^{(3)} = 2 \times 10^{-11}$ esu. These values present a difference of up to three orders of magnitude compared to lithium niobate crystals exhibiting a third-order nonlinear optical susceptibility of $\chi^{(3)} = 4.3 \times 10^{-11}$ esu [7]. The nonlinear optical response exhibited by films deposited on different substrates has shown promise for nonlinear photonic applications. Additionally, the saturable absorption exhibited by lithium niobate crystals can be used for ultrafast optical modulation in laser oscillators to generate ultrashort laser pulses.

Correspondingly, with the combination of organic materials with metallic nanoparticles, the nonlinear optical properties of the materials present convenient improvements. One of the interesting semi-organic materials among those derived from the combination with metallic nanoparticles is hydrogen maleate dihydrate. Lithium hydrogen maleate dihydrate crystal exhibits good nonlinear optical responses. Thus, the third-order nonlinear optical susceptibility of lithium hydrogen maleate dihydrate has a value of $\chi^{(3)} = 2.76 \times 10^{-6}$ esu [8]. The third-order nonlinear response is comparably higher than other lithium-based semi-organic crystals.

The generation of harmonics through laser pulses is an efficient and important technique in the study of the optical nonlinearities of semi-organic crystals. Through the generation of the third harmonic, it is possible to study nonlinear optical processes in crystals through third-order nonlinear optical susceptibility. So, pure lithium triborate, which is also a nonlinear optical crystal, has been studied through the generation of the third harmonic at 532 nm. Thus, the third-order nonlinear optical susceptibility of pure lithium triborate exhibits a value of $\chi^{(3)} = 6.75 \times 10^{-5}$ esu [9].

6.3   Sodium Crystals Second- and Third-Order Optical Nonlinearities

Generally, nonlinear optical susceptibilities are located in materials that possess π-conjugated molecules, together with an expanded column of delocalized electrons. Thus, π-electron cloud propagation within these organic materials is the main cause of third-order nonlinear optical responses. At the time, due to the chemical structure, crystals can exhibit third-order nonlinear optical susceptibilities.

Thus, sodium niobate nanocrystals have been synthesized using the sol–gel method in order to perform second-order optical nonlinearity studies. Also, the first hyperpolarizability of the nanoparticles was measured using a laser operated at a wavelength of 1064 nm. Using the hyper-Rayleigh scattering technique, the scattered light of the second harmonic generated by the sodium niobate nanocrystals was analyzed. Thus, the hyperpolarizability value of sodium niobate per volume of nanoparticles has been determined to be 0.93×10^{-29} esu/nm^3. On the other hand, for nanocrystals with a mean diameter of 109 nm, the first hyperpolarizability exhibited a value of 6.3×10^{-24} esu [10].

The nonlinear optical properties of sodium tantalum germanate glasses have been investigated through the second harmonic generation. Thus, the value of the second-order optical susceptibility $\chi^{(2)}$ ranges between 0.06 and 0.3 pmV^{-1}. Moreover, through the second harmonic generation, structural changes have been identified in the bonds of sodium tantalum germanate [11].

Furthermore, the result of the thermal effects and the addition of sodium in the efficiency and stability of the study of second-order nonlinearities in amorphous tantalum pentoxide films have been investigated. The study was carried out in analysis of the second harmonic generation. Thus, the second-order optical susceptibility result in $\chi^{(2)}$ corresponds to a value of 0.34 pmV^{-1} for amorphous tantalum pentoxide films containing sodium. Thus, the response of the second harmonic generation is associated with the density of the films [12].

Thus, the optical properties of glycine sodium nitrate single crystal have exhibited a photoconductive nature. Also, photoconductivity increases with respect to temperature. Then, the third-order nonlinear optical response of the glycine sodium nitrate single crystal has shown a value of $\chi^{(3)} = 4.34 \times 10^{-5}$ esu [13]. These properties exhibited by single-crystal glycine sodium nitrate demonstrate that it is a good demanding material for application in nonlinear photoconductive optical devices.

On the other hand, ultrafast third-order nonlinear optical properties of sodium borate have been studied. Studies have shown that sodium borate compounds possess optical clarity and a high linear refractive index. The nonlinear optical response of sodium borate compounds at an antimony concentration of 10 mol% and at wavelengths of 750, 800, and 880 nm have exhibited values of $\chi^{(3)} = 2.87 \times 10^{-13}$ esu, $\chi^{(3)} = 1.6 \times 10^{-13}$ esu, and $\chi^{(3)} = 1.2 \times 10^{-13}$ esu, respectively. Besides, the responses at a concentration of 20 mol% show values of $\chi^{(3)} = 5.46 \times 10^{-13}$ esu, $\chi^{(3)} = 4.54 \times 10^{-13}$ esu, and $\chi^{(3)} = 4.19 \times 10^{-13}$ esu at wavelengths of 750, 800, and 880 nm. On the other hand, at a concentration of 20 mol% they exhibit values of $\chi^{(3)} = 10.1 \times 10^{-13}$ esu, $\chi^{(3)} = 8.83 \times 10^{-13}$ esu, and $\chi^{(3)} = 7.72 \times 10^{-13}$ esu at the wavelengths mentioned previously [14]. Nonlinear susceptibility values for sodium borate at different concentrations and wavelengths have shown that they are potential materials for limiting optical applications.

On the other hand, inorganic crystals have chemical advantages that provide high melting points, thermal stability, and remarkable mechanical behavior, unlike totally organic crystals that have weak Van der Waals forces. Also, inorganic crystals demonstrate optical clarity, fast optical response time, and wide-phase adaptation angle. Moreover, the third-order nonlinear optical property has been studied for the developed inorganic crystal of sodium cadmium tetrachloride and the results indicate that the crystal is a promising material for nonlinear optical device applications. The nonlinear third-order optical susceptibility measured was $\chi^{(3)} = 4.32 \times 10^{-8}$ esu [15].

Moreover, third-order optical nonlinearities of inorganic single crystals of sodium penta borate grown at wavelengths of 532 nm have been studied. The third-order nonlinear optical response of sodium penta borate single crystals presented a value of $\chi^{(3)} = 2.07 \times 10^{-7}$ esu [16]. The nonlinear refractive index of sodium penta borate single crystals depends on the intensity of the incident beam, which results in self-diffraction in the incident beams. The self-diffraction effect in these materials is related to nonlinearities derived from thermo-optical mechanisms and is useful for applications in optical limiting devices.

6.4 Potassium Crystals Second- and Third-Order Optical Nonlinearities

Materials that exhibit ultrafast optical responses produced in the nano, peak, and femtosecond optical signal regime are extremely important for the development of nonlinear photonic applications. Single crystals have attracted attention due to their morphology, in which the crystal lattice is continuous and not interrupted by grain edges up to the limits of the sample. The geometric structure of these materials has carried out characterization and optical response studies of the single crystals.

The influence of thermal effects on the properties of the second generation of harmonics in potassium oxide glass has been studied. It has also been analyzed that second-order nonlinearities of potassium oxide are associated with superficial structural modifications. In addition, the structural changes result from a charge transport process that results in the modification of the network in an alkali-depleted layer whose thickness is comparable to that of the nonlinear region. Thus, the second-order optical susceptibility value in potassium oxide glasses is 3.8 pmV^{-1} [17].

On the other hand, the crystalline phases have also been studied through the generation of the second in crystallized potassium oxide glasses. It is also established that the behavior of the crystalline structure depends on the molecular compound of potassium that composes it. Thus, through the formation of stable crystalline phases, several types of metastable crystalline phases are formed that show a good nonlinear response to the second harmonic generation [18].

On the other hand, the second-order nonlinear response in potassium niobium and potassium silicate glasses has also been studied when exposed to prolonged heat treatments in a wide range of temperatures. The heat treatments result in the separation of liquid-type phases in potassium glasses. Furthermore, for each glass composition, temperature zones have been determined to produce transparent, opalescent, or opaque materials. In addition, it has been determined that nanostructured potassium glasses exhibit nonlinear activity derived from the second harmonic generation. However, the increase in opalescence in the crystalline phases leads to a decrease in response in the second harmonic generation. This nonlinear behavior before the second harmonic generation of potassium glasses is related to combined mechanisms of second-order optical nonlinearity and third-order nonlinearity with the spatial modulation of linear polarization [19].

On the other hand, in another study the response of second-order optical nonlinearities was measured through the application of the hyper-Rayleigh scattering technique on titanyl potassium phosphate particles such that the first-order optical hyperpolarizability exhibited a magnitude of 33.5×10^{-24} esu. In addition, in potassium niobate particles a value of 21.8×10^{-24} esu was exhibited [5].

Single crystals of potassium thiourea chloride have shown a high contribution to nonlinear optical susceptibility of the third-order magnitude $\chi^{(3)} = 8.71 \times 10^{-5}$ esu [20]. The high value in third-order nonlinear optical susceptibility is associated with the delocalization of the photoinduced charge on the links π. The nonlinearities of single crystals of potassium thiourea chloride show significant improvements compared to other metallic components of thiourea. In addition to presenting a high ablation threshold due to its continuous geometry, the effects that dominate the third-order nonlinear optical response of single crystals are nonlinear refraction and absorption. Characterization and nonlinearity studies show that single crystals of potassium thiourea chloride are attractive for applications related to night-vision sensors, optical limiters, and photonic devices.

On the other hand, in the family of semi-organic borates are the single crystals of potassium borodicitrate. This family of semi-organic single crystals is remarkable due to the donor-pacifier configuration (D-p-A). This configuration is associated with materials with high optical nonlinearity responses. Also, single crystals of potassium borodicitrate have a configuration in which the bicitrate borate anions act as electron receptors for the central atom.

The crystalline structure of potassium borodicitrate also favors the photoconductive nature of the material. Therefore, the third-order nonlinear susceptibility of the potassium borodicitrate single crystal is $\chi^{(3)} = 4.1 \times 10^{-9}$ esu [21]. Furthermore, the laser-induced ablation threshold for the crystal demonstrated the thermal resistance of the material.

The potassium hydrogen phthalate crystal possesses high nonlinear optical properties due to the strong interactions of the hydrogen bonds of the polarizable aromatic rings of the phthalate ions. The third-order optical susceptibility response of potassium hydrogen phthalate crystals established a value of $\chi^{(3)} = 1.95 \times 10^{-6}$ esu [22]. A comparison with other crystal structures reveals that this material has a good third-order nonlinear optical response. Unlike inorganic crystals, organic crystals with nonlinear optical characteristics possess a high nonlinear and electro-optical optical coefficient, low dielectric dispersion, ultrafast response time, and highly adaptive structure. Furthermore, crystalline organic materials have the advantage of controlling their structures due to the modification of their nonlinear optical properties. Inorganic potassium crystals are often widely used in applications related to frequency conversion due to their excellent nonlinearities. On the other hand, L-leucine potassium chloride is a widely used nonlinear optical response amino acid. Amino acids are considered as zwitterions since they contain a group of COOH proton-donating carboxylic acids and a proton-accepting NH_2 group. Furthermore, the crystals of potassium chloride L-leucine have a monoclinic structure with a crystalline nature. Thus, third-order nonlinear optical susceptibility exhibits a magnitude of $\chi^{(3)} = 1.95 \times 10^{-6}$ esu [23] due to the self-focusing nature of the sample. Such a parameter shows that the crystal of potassium chloride L-leucine is a material with potential characteristics for nonlinear optical applications.

On the other hand, the crystal of zinc, potassium, and aluminum sulfate nonadecahydrate presents a third-order nonlinear susceptibility value $\chi^{(3)} = 2.13 \times 10^{-6}$ esu. The crystal shows high third-order nonlinear optical susceptibility due to highly polarized laser beam irradiation [24].

6.5 Cesium Crystals Second- and Third-Order Optical Nonlinearities

Nowadays, several photonic applications are based on nonlinear optics, which implies very high optical irradiances and therefore high-intensity lasers. Thus, it is necessary to study sensitive optical materials and high ablation thresholds. Due to this, attention has been paid to the study of third-order nonlinear optical parameters and optical limiting properties of crystals. Besides, the study of crystal growth parameters is focused on to define the control of optical modifications due to the geometry of the crystals.

The second harmonic generation of critically phase unpaired harmonics in lithium cesium borate has been shown to provide high efficiency at lower voltages using high peak power lasers, providing longer crystal life [25].

With respect to cesium lithium borate crystals, the optical coefficients of second-order nonlinearities based on the molar ratio at different compositions have been measured. Besides, two-phase pairing configurations have been studied for the two crystal compositions through the second harmonic generation such that a magnitude of 0.46 and 0.92 pmV^{-1} has been exhibited for each of the molar compositions [26].

Metal halide perovskites is a hybrid material that has exhibited great potential for the development of photonic applications. Metal halide perovskites in their crystalline form demonstrate predominant optical gains under the incidence of fluorescent light. Furthermore, the photoluminescent nature of the material demonstrates that the bandgap and perovskite emission spectra are manipulable through the visible electromagnetic spectrum. Furthermore, the colloidal nanocomposites of cesium lead halide perovskite nanocrystals are very characteristic due to their high coefficient of optical absorption. Then, colloidal nanocomposites have been studied using different cesium lead halide compound methods in order to determine whether the anion exchange reactions and recrystallization due to thermal effects establish modifications in the nonlinear optical response. Thus, for colloidal cesium lead halide compounds nonlinear refraction is the main third-order optical nonlinearity response. The nonlinear third-order optical susceptibility of colloidal compounds of $CsPbCl_3$ perovskites exhibits a value of $\chi^{(3)} = 2.85 \times 10^{-11}$ esu. On the other hand, the third-order nonlinear optical susceptibility of colloidal compounds of $CsPbBr_3$ perovskites exhibits a value of $\chi^{(3)} = 3.78 \times 10^{-11}$ esu. Moreover, the third-order nonlinear optical susceptibility of colloidal compounds of $CsPbI_3$ perovskites shows a value of $\chi^{(3)} = 3.55 \times 10^{-11}$ esu [27].

On the other hand, cesium sulfate is an inorganic compound that is frequently used to prepare colloids. The estimated third-order susceptibility value of cesium sulfamate is $\chi^{(3)} = 9.26 \times 10^{-8}$ esu [28]. Optical studies performed on cesium sulfate reveal that the material has a positive refractive index and results from the self-diffraction nature. Also, the material exhibits the phenomenon of absorption of two photons of the crystal. This shows that the optical parameters of cesium sulfamate are suitable for optical limiting applications. In contrast, organic metal structures are crystalline porous hybrid materials. The single crystal of cesium structure extended with oxalate bridges possesses favorable third-order nonlinearities for the development of applications based on the confinement of optical power, storage of optical and quantum information, and photodynamic. The third-order nonlinear susceptibility of the cesium hydrogen oxalate dihydrate crystal is $\chi^{(3)} = 5.68 \times 10^{-6}$ esu [29].

For cesium tetroxalate dihydrate crystal, the exhibited high third-order nonlinear susceptibility and low threshold for optical limitation establish the material as a promising candidate for optical limitation. Thus, the third-order nonlinear optical response of cesium tetroxalate dihydrate crystal exhibits a magnitude of $\chi^{(3)} = 5.68 \times 10^{-6}$ esu [30]. Furthermore, third-order nonlinear optical parameters are modified due to the structural, spectroscopic, thermal, mechanical, and dielectric properties of the crystals.

6.6 Rubidium Crystals Second- and Third-Order Optical Nonlinearities

Crystalline nonlinear optical materials are used for the development of solid-state laser sources. That is why the crystals chosen to form the sources need to satisfy intrinsic and extrinsic properties. This is in addition to the fact that it is necessary that they should be a material with manipulable properties since it is also necessary to establish control over the optical properties of the glass. In crystals used for frequency conversion, it is necessary for the crystal to operate at discrete wavelengths and move through the electromagnetic spectrum ranging from ultraviolet to far-infrared. Therefore, it is necessary to combine a range of materials and synthesis techniques in order to develop crystals with operation in the various wavelength regimes. In the frequency conversion process derived from a nonlinear optical material, it is necessary to enter a laser beam at one or two wavelengths through a crystal. That proceeds to the generation of laser light in one or two different wavelengths at the exit of the crystal. This interaction through crystals with a frequency conversion motive can be understood as a coupling of optical waves, and it is necessary to conserve the energy of the optical beams.

Moreover, extensive efforts have been made to design new nonlinear optical materials with the property of response to the second harmonic generation. This is to develop semi-organic crystalline nanomaterials with utility in applications such as frequency conversion, light amplitude, and phase modulation and phase conjugation due to their great nonlinearity. The contribution of delocalized electrons from the organic ligand results in increased nonlinear optical and electro-optical behavior in semi-organic materials. Thus, semi-organic nonlinear optical rubidium nanocrystals have been systemized from cooling. Also, the response of the second harmonic generation through an Nd:YAG laser has been confirmed [31].

Rubidium titanyl phosphate is a nonlinear crystal that has large nonlinear optical coefficients. The electro-optic study in these materials has been employed for the measurement of second-order nonlinearities through nonlinear phase shift. It has been pointed out that it is possible to generate a second-order nonlinearity in cascading the rubidium phosphate nanocrystals such that the effective nonlinear refractive index has a value of 1.4053×10^{-13} cm^2/W [32].

Crystals composed of titanyl phosphate with rubidium substitutions have been investigated. The crystal has exhibited high electrical resistivity and nonlinearities that make it attractive for use as an electro-optical switch. On the other hand, rubidium titanyl arsenate exhibits high resistivity and transparency in infrared regions. Also, the nonlinearities of binary borate crystals at different concentrations of rubidium have been studied. The nonlinear optical response of binary borate crystals at rubidium concentration of 10 mol% exhibited values of $\chi^{(3)} = 2.99 \times 10^{-14}$ esu. Besides, the responses at a concentration of 20 mol% show values of $\chi^{(3)} = 3.10 \times 10^{-14}$ esu. On the other hand, at a concentration of 30 mol% they exhibit values of $\chi^{(3)} = 3.74 \times 10^{-14}$ esu [33].

Compounds of binary borate crystals have shown an increase in third-order nonlinear optical response to increasing rubidium concentration in the crystal. The third-order nonlinear optical responses of binary borate crystals estimated through the generation of the third harmonic have found a relationship with the crystal structure. The improvement in third-order nonlinear optical susceptibility is manifested due to the presence of nonbridging oxygens above 20 mol% rubidium. The lithium rubidium borate crystal is a potential crystalline material with nonlinear optical characteristics and belongs to the alkali metal borate family. The lithium rubidium borate crystal structure contains two types of anionic groups. This crystal is very characteristic due to the crystalline perfection of its structure and the optical homogeneity of its properties. The third-order nonlinear optical susceptibility of the lithium rubidium borate crystal is estimated to be $\chi^{(3)} = 2.71 \times 10^{-7}$ esu [34].

On the other hand, the third-order nonlinear optical susceptibility response of rubidium has shown a significant improvement through the optical wave coupling process. Equally, the improvement in nonlinearities is associated with the electromagnetic transparencies induced on a rubidium compound. The induced transparencies lead to the efficiency of the nonlinearities derived from the coupling of optical waves because a reduction in optical absorption is generated in rubidium crystals from the resonance of light through the generated quantum phenomenon. As a result of the reduction in absorption by 90% of the generated conjugate wave, it is possible to consider the generation of compressed light through the coupling of optical waves [35].

6.7 Complex Geometries

The optical properties dependent on the size and shape of nanoparticles have provided the advantage of being able to design materials with controllable characteristics with wide potential applications. Thus, the design and synthesis of structurally complex nanoparticles improve the functionality of their physicochemical characteristics. The synthesis of nanoellipsoids, nanostars, nanoflowers, nanoribbons, and other complex nanostructures can be highlighted. In addition, they can be complemented by the incorporation of nanoparticles of other elements to design hybrid nanoparticles with improved properties. An example of this is the synthesis of semiconductor nanowires with a coated structure that allows electronic interactions within the nanowire. In addition, the Cu and Pt nanowires also exhibit tunable magnetic properties. Such materials can exert their properties as nanophotonic light-emitting diodes. On the other hand, quantum-type applications and information processing are achieved through the synthesis of branched quantum dots. As is known, metallic nanoparticles exhibit localized surface plasmon resonances, resulting in strong optical extinction at visible wavelengths. The localized surface plasmon resonance of metallic nanoparticles can be modified through the design of geometrically complex metallic nanostructures. As an example of the advantages of geometry, gold nanostars can modify the optical extinction response at near-infrared wavelengths, enabling important biomedical applications such as biological and chemical detection, biological imaging labels, and nano-optical waveguides. In addition, the optical properties of metallic nanoparticles with star geometries incorporate polarization-dependent dispersion properties with multiple spectral peaks and strong dielectric sensitivity in a single structure.

6.8 Nanoellipsoids Second- and Third-Order Optical Nonlinearities

Second harmonic generation from gold nanoellipsoids is investigated using high-power femtosecond polarized light pulses. Through this, it is found that the nonlinear near-field response is directly related to both the local surface plasmon resonances and the morphology of the nanoparticles. Thus, the high maximum power light pulses in the near field allow the local excitation of emission of the second harmonic generation of nanoellipsoids in a size range smaller than 100 nm [36].

On the other hand, second harmonic generation has been used to determine the field enhancements of Au nanoellipsoids on a silicon substrate in which the nonlinear optical spectra have been found to be dominated by a horizontal plasmon resonance close to 1.0 eV. Furthermore, through the second harmonic generation, resonances of the nanoellipsoids and the silicon oxide substrate are observed. Besides, the combination of charge transfer gives rise to second-order nonlinearities and an improvement in the optical field in a thin surface layer of the Si substrate [37].

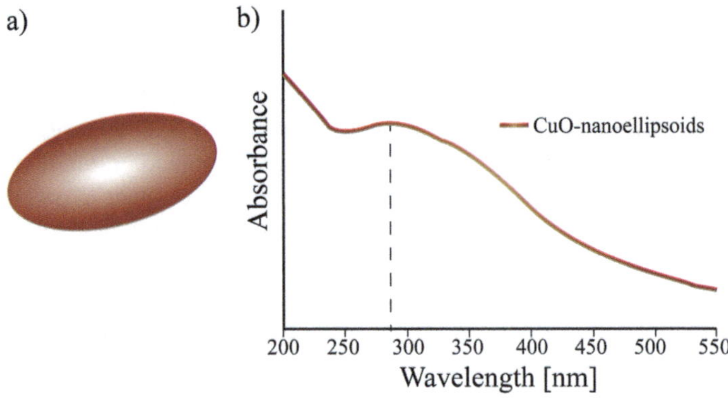

Fig. 6.2 (**a**) Geometrical representation of CuO nanoellipsoids; (**b**) UV–VIS spectrum CuO nanoellipsoids

The third-order optical susceptibility response of suspended copper oxide nanoellipsoids described by UV–VIS data in Fig. 6.2, for a pulse irradiation of 60 fs and wavelengths at 400 nm and 600 nm, has been studied. A magnitude in third-order nonlinear optical susceptibility of $\chi^{(3)} = 2.8 \times 10^{-9}$ esu was demonstrated at a wavelength of 400 nm. In another way, third-order nonlinear optical susceptibility exhibited a value of $\chi^{(3)} = 1.2 \times 10^{-9}$ esu at a wavelength of 600 nm [38].

In relation to the third-order nonlinear optical response, it is necessary to highlight that quantum confinement effects are present due to the morphological characteristics of the copper oxide nanoellipsoids. Furthermore, there is a dependence of the optical properties of the material with the synthetic surface area of the materials and the volume of the medium that supports the copper oxide nanoellipsoids. Thus, the relationship of the optical nonlinearities of the ellipsoids is inverse to the decrease in the magnitude of the radius of the ellipsoids. Besides, the improvement in third-order optical nonlinearity can be defined due to the concentration of the carriers on the surfaces of the copper oxide nanoellipsoids. Moreover, in the specific case of copper oxide nanoellipsoid suspensions, it has been determined that they have positive nonlinear refractive properties, derived from ultrafast pulses of fs.

Correspondingly, the nonlinear optical response of mechanically stretched metallic nanoellipsoids has shown a value of $\chi^{(3)} = 4.9 \times 10^{-10}$ esu at a wavelength of 1030 nm [39]. On the other hand, the nonlinear optical response of mechanically stretched metallic nanoellipsoids has shown a good third-order nonlinear optical response due to the geometric anisotropy of the nanoellipsoids. The mechanism that gives way to the third-order nonlinear optical response of metallic nanoellipsoids is established through the contribution of hot electrons from the metal. Furthermore, the third-order nonlinear optical response is also associated with the localized surface plasmon resonance response of the nanoparticles. The response of the plasmonic resonance for metallic nanoellipsoids is affected at wavelengths close to the infrared region of the electromagnetic spectrum and also through polarization rotation of the beam at an angle of $90°$. Similarly, titania semiconductor nanoellipsoids have optical properties that allow applications and have the advantage that it is possible to prepare them in polymeric, ceramic, and glass matrices. Many of the optical applications for which titania ellipsoids are used include optical coatings, photocatalysts, and sensors because such nanoparticles allow for size-dependent bandgap change, high carrier mobility, and nonlinear optical properties. Moreover, inorganic titania nanoellipsoids due to their size and morphology allow their dispersion in a transparent optical polymer of polymethyl methacrylate. This polymer matrix has the advantage of being versatile and suitable for studying the optical properties of nanostructures.

The nonlinear optical response of titania ellipsoid films embedded in polymethyl methacrylate matrices was studied due to the percentage of the weight contained in the films such that films with a weight% of titania nanoellipsoids exhibit a value in third-order nonlinear optical susceptibility of $\chi^{(3)} = 0.4 \times 10^{-9}$ esu. Titania-based films with 40% by weight show $\chi^{(3)} = 1.7 \times 10^{-9}$ esu. Also, the third-order nonlinear optical susceptibility of the 60% weight films $\chi^{(3)} = 3.6 \times 10^{-9}$ esu. Finally, films with 80% of the weight exhibit nonlinearities with a magnitude of $\chi^{(3)} = 0.6 \times 10^{-9}$ esu [40]. The improvements in the nonlinearities of the films are due to the fact that the nanoellipsoids of titania have a high dielectric confinement effect, while the quantum confinement effects of this nanomaterial are weak. Furthermore, bismuthate glasses embedded with gold ellipsoids have demonstrated low-pulse high-order nonlinearities in the fs regime. The nanocomposites exhibited a third-order nonlinear susceptibility of $\chi^{(3)} = 4.88 \times 10^{-10}$ esu to pulses of wavelength 800 nm [41]. The nonlinearities demonstrated by the gold ellipsoid compounds have been shown to be improved due to the compound's intraband transition mechanism.

6.9 Nanorods Second- and Third-Order Optical Nonlinearities

The response of the second-order nonlinearities of nanoparticles depends on the size and shape. Thus, the optical nonlinearities of gold nanorods have been studied through the hyper-Rayleigh scattering technique. The nanorods studied have an average aspect ratio of length = 38 nm and width = 17 nm. Furthermore, for the measurements of the first hyperpolarizability of the nanorods, the densities of the sample were measured. The numerical densities of the nanorods were estimated from their sizes such that the magnitude of the first hyperolarizability associated with gold nanorods is 41×10^{-25} esu [42].

On the other hand, the optical nonlinearities of nanorod particle matrices excited by femtosecond laser pulses have been studied from the result of the surface resonance plasmon that originates due to the second harmonic generation. These measurements have been analyzed as a function of the polarization of the incident light and the wavelength of the irradiation. Besides, there is a phenomenon of

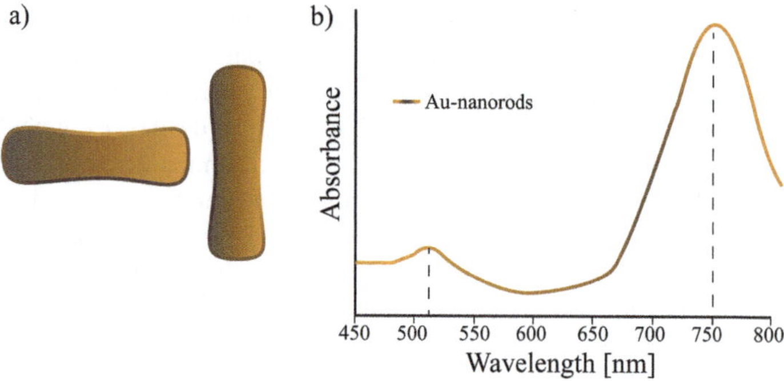

Fig. 6.3 (**a**) Geometrical representation of Au nanorods; (**b**) UV–VIS spectrum Au nanorods

photoluminescence derived from the second harmonic generation, and it is found that it depends on the polarization and the wavelength of the beam [43].

On the other hand, the dynamics of nonlinear second-order fields scattered through a single nanorod irradiated by a plane-wave beam of femtosecond pulses has been studied. Through this, it has been observed that the transverse dipole mode interferes with the quadrupole mode at the time of the excitation pulse. Furthermore, different second-order nonlinearity responses have been associated as a function of the beam pulse width. Furthermore, it has been established that the structure of the nanorod is of importance in understanding the temporal evolution of nonlinear fields [44].

It has also been shown that the interaction between gold nanorods featuring an optical spectrum as shown in Fig. 6.3 and a strong electric field produces a confinement and causes the improvement of the optical nonlinearities exhibited by the material. The nonlinear third-order optical susceptibility of gold nanorods films has shown a magnitude of $\chi^{(3)} = 4.7 \times 10^{-9}$ esu to pulses with wavelength of 740 nm. Therefore, at 780 nm wavelength pulses it exhibits $\chi^{(3)} = 7.2 \times 10^{-9}$ esu. Also, at a wavelength of 800 nm, the susceptibility expresses a value of $\chi^{(3)} = 6.0 \times 10^{-9}$ esu. Finally, the third-order nonlinearities exhibited at a wavelength of 820 nm demonstrate a magnitude of $\chi^{(3)} = 5.7 \times 10^{-9}$ esu [45].

Au nanorods have also been shown to exhibit improved third-order optical nonlinearities unlike bulk Au; the above due to its morphological conditions. The nonlinear optical response of metallic nanomaterials is strongly influenced by local field effects in addition to wavelength dependence. The third-order nonlinear optical susceptibility of Au nanorods demonstrates a value of $\chi^{(3)} = 1.0 \times 10^{-10}$ esu [46]. The response of the nonlinearities of Au nanorods determines a nonlinear absorption and refraction and dependence on the surface plasmon.

Besides, Ag-coated Au nanorods possess a broad wavelength tuning capacity due to the surface plasmon resonance response exhibited by metals. Nonlinear third-order optical properties of coated nanorods with different thicknesses have been investigated, which have demonstrated a maximum third-order nonlinear optical susceptibility of $\chi^{(3)} = 109 \times 10^{-11}$ esu [47]. The resulting nonlinearities of the nanorods at different thickness demonstrate that the surface plasmon resonance response of the nanorods is sensitive to the thickness of the Ag coating and the change in the dielectric constant. This indicates that the wavelength-dependent third-order optical susceptibility is in turn modified due to localized surface plasmon resonance changes.

6.10   Nanoribbons Second- and Third-Order Optical Nonlinearities

The presence of π-electrons in graphene derivatives can generate structures with large values of the first optical hyperpolarizability in cases where the centrosymmetry is dressed. Such is the case of graphene nanoribbons, in which a first hyperpolarizability is exhibited with a magnitude of $220,000 \times 10^{-31}$ esu [48].

Furthermore, the optical nonlinearities in zinc oxide nanoribbons have been studied through high-density samples over a large area. Thus, the results of the measurements of the second-order nonlinearities suggest large nonlinear effects associated with matrices of zinc oxide nanoribbons. Thus, a possible application of zinc oxide nanoribbon matrices as optoelectronic nanodevices is established [49].

Graphene nanoribbons with optical absorption described by Fig. 6.4 possess a remarkably good third-order nonlinear optical response. This is due to its response in the long wavelength regime and in a wide frequency range. Nanoribbons have been shown to possess strong third-order Kerr susceptibility and third-order susceptibility with a magnitude of $\chi^{(3)} = 7 \times 10^{-7}$ esu [50].

The third-order nonlinear optical susceptibility exhibited by graphene nanoribbons reduces the intensity of the optical field, making the material promising for photonic applications. Graphene nanoribbons attract interest due to their electron confining properties and for applications in optoelectronic devices. In particular, the nonlinear optical response of graphene nanoribbons is dominated by pairs of electrons and holes, excitons. The third-order nonlinear optical response $\chi^{(3)}$ of graphene nanoribbons shows values around the order of 1×10^{-7} esu [51]. The response of

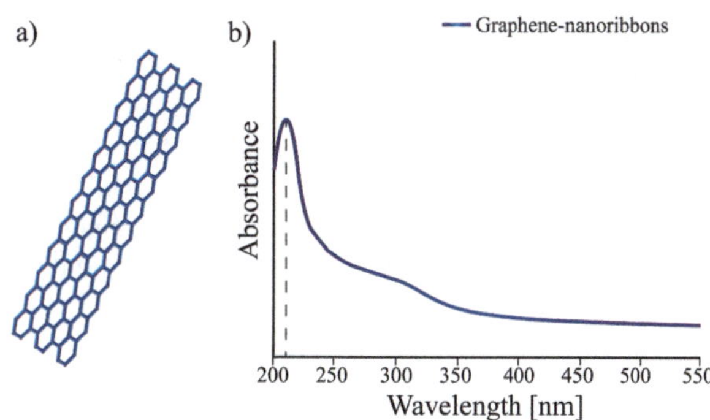

Fig. 6.4 (a) Geometrical representation of graphene nanoribbons; (b) UV–VIS spectrum graphene nanoribbons

third-order optical susceptibilities depends on the geometry of the slats, such that nonlinearities increase due to the increase in the radius and width of the nanostructures; this is also due to the bandgap of the material.

6.11 Nanowires Second- and Third-Order Optical Nonlinearities

On the other hand, beryllium nanowires have been studied based on their optical nonlinearities. Such structures have been shown to exhibit considerable second-order nonlinear optical responses for application design. Furthermore, nanowires show the electronic properties with stronger electron-donating and electron-withdrawing behaviors and possess wave bands in the infrared and ultraviolet regions. This is associated with the effect of the doping of atoms of other elements that produce an increase in the value of the first optical hyperpolarizability. Furthermore, it is notable that there is an effect due to the length of the nanowire on the value of the first hyperpolarizability, which acts to increase gradually [52].

On the other hand, nanowires synthesized from lead halide are included in the classification of non-centrosymmetric materials, for which they demonstrate a highly efficient second harmonic generation with high polarization ratios and chiroptic second-order nonlinear optical effects [53].

In reference to nanomaterials with other forms, the anisotropy associated with zinc oxide nanowires is attributed to the contribution of third-order nonlinearity due to the crystal lattice of the nanowires, in addition to the contribution of the quantum confinement effect. Also, zinc oxide nanowires define a large two-photon absorption coefficient and nonlinear refractive index modification. Correspondingly, zinc oxide nanowires have been observed to exhibit oscillation patterns of two-photon absorption and nonlinear refraction due to the orientation of the polarization angle of the incident fs beam. Furthermore, there is a dependence of the size of the nanowires with the optical nonlinearities of the samples [54].

Moreover, metallic nanowires possess positive refractive nonlinearities and inverse saturated absorption behaviors. The nonlinear optical susceptibility of Ag and Cu metallic nanowires with absorption band described in Fig. 6.5 has been studied through a wavelength of 532 nm in an 8 ns pulse regime. The magnitude of the third-order nonlinear optical susceptibility of the samples shows $\chi^{(3)} = 1.7 \times 10^{-11}$ esu for Ag nanowires; it is also demonstrated that $\chi^{(3)} = 2.4 \times 10^{-11}$ esu for samples of Cu nanowires [55]. The nonlinearities associated with nonlinear refraction are defined through self-diffraction induced by the optical Kerr effect since the thermal effect does not have much influence on the value of the refractive indexes of the samples.

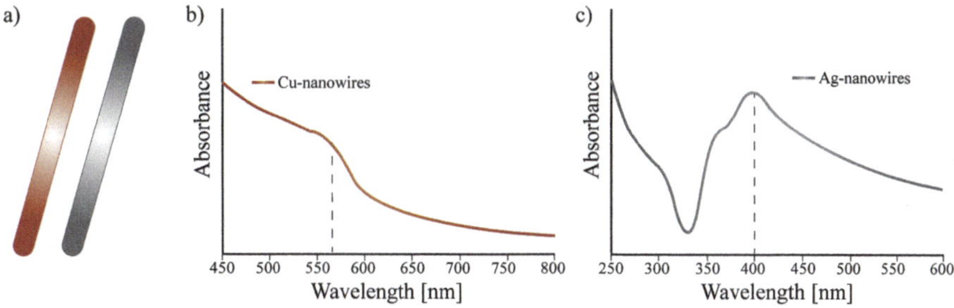

Fig. 6.5 (**a**) Geometrical representation of nanowires; (**b**) UV–VIS spectrum Cu nanowires; (**c**) UV–VIS spectrum Ag nanowires

6.12 Nanoflowers Second- and Third-Order Optical Nonlinearities

Nanoflowers are nanometric structures that, due to their size and morphology, have a high conductive capacity. Due to this, they are considered potential materials in the design and application of energy storage devices. Otherwise, these nanostructures also have optical limitation characteristics, and the properties demonstrate an induced decrease in transmittance at very high energies, in addition to excellent third-order nonlinear optical response.

Regarding the response of second-order optical nonlinearities, flower-shaped gold nanoparticles have a higher hyperpolarizability, up to 52 times greater than other forms of gold nanoparticles. These improvements are attributed to reduced symmetry such that the magnitude of the first hyperolarizability associated with gold nanoflowers is 31×10^{-25} esu [42].

Nanoflowers have demonstrated optical characteristics based on efficient nonlinear absorption mechanisms. Such mechanisms involve reverse saturable absorption, multiphoton absorption, and the free carrier absorption mechanism. Vanadium oxide nanoflowers have been shown to possess optical properties based on three-photon absorption mechanisms. Furthermore, vanadium oxide nanoflowers have also been observed to possess ferromagnetic properties at a temperature of approximately 26 °C [56].

On the other hand, MoS_2 nanoflowers with optical absorption shown in Fig. 6.6 exhibit interesting nonlinear optical properties, such as photoluminescence, as well as responses to second- and third-order susceptibility, and ultrafast nonlinear optical absorption. Due to this, some research based on the exploration of nonlinear optical behavior has focused on nanoflowers. In addition, it has been established that there is a relationship between the nonlinearities of the material with the thickness of the nanoflowers. Evenly, nanoflowers with sizes greater than 100 nm demonstrate saturable absorption responses, unlike MoS_2 nanoflowers with sizes less than 50 nm that demonstrate nonlinear inverse saturable absorption responses.

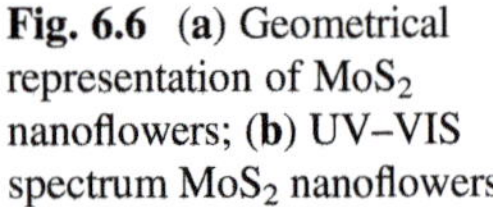

Fig. 6.6 (**a**) Geometrical representation of MoS₂ nanoflowers; (**b**) UV–VIS spectrum MoS₂ nanoflowers

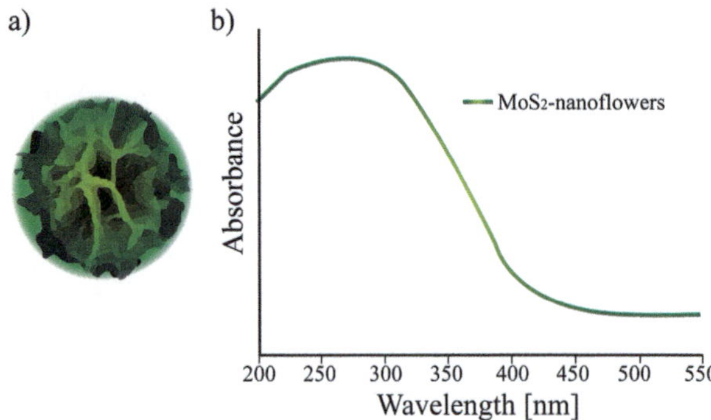

Derived from optical studies performed on MoS₂ nanoflowers, a significant saturable absorption response has been observed. Furthermore, the refractive index exhibited by MoS₂ nanoflower samples has been estimated at $n_2 = 3.9 \times 10^{-15}$ cm²/W [57]. This good nonlinear optical response in the MoS₂ nanoflowers is important because they are excellent candidates for photonic and photoelectronic applications.

Otherwise, nanoflowers of tin sulfide–cadmium sulfide, due to their optical properties, can be used in optical switches based on saturable absorption mechanisms to define the modulation of light periodically. The saturable absorption mechanism arises from third-order optical nonlinear absorption interactions of nanoflowers [58].

6.13 Nanoplanets Third-Order Optical Nonlinearities

Nanoplanets are a type of nanoparticles that consist of nanomaterials with a central core surrounded by small groups of satellites. The nanoplanets are also called core satellite nanoparticles.

Nanoplanets have been shown to have great plasmonic and optical properties due to local field enhancements in the halos of the satellite clusters around the core of the nanoparticle that make up the nanoplanets.

On the study of the properties of nanoplanets, it is determined that it is important to analyze them through an electromagnetic interaction. The plasmonic properties of Au–Ag alloy nanoplanets depend on the conditions of the light beam irradiated on the material; it has been determined that the plasmon of nanoplanets occurs close to the red region of the visible electromagnetic spectrum. Furthermore, the plasmonic properties of nanoplanets also depend on the property that clusters of nanoparticles defined as satellites have to remain unaltered in the face of the interacting beam. Furthermore, it is possible to control the optical and plasmonic properties of Au–Ag alloy nanoplanets up to 60 nm. The modification of the resonance plasmon response is due to variations in the local dielectric environment of the nanoplanet nucleus, in addition to the coupling of the adjacent satellite cluster [59].

Due to the characteristics described, it is necessary to affirm that the metallic nanoplanets of Au–Ag alloys are a promising type of nanomaterial in the field of nonlinear optical applications since the third-order nonlinear response has a great relationship with the response plasmonics of nanomaterials. Plasmonic characteristics influence the absorption behavior of Au–Ag nanoplanets, resulting in the conversion of inverse saturable absorption to saturable absorption. This is achieved through the modification of the beam intensity parameters. This also has the consequence of a decrease in the absorption coefficient depending on the intensity of the beam, in the specific case of Au–Ag nanoplanets. In general, local field effects produced on nanoplanets at wavelengths susceptible to surface plasmon resonance enhance the nonlinear optical response of the nanomaterial.

The nonlinear optical characteristics of Au–Ag nanoplanets have been studied in the ps regime at wavelengths close to the surface plasmon resonance of the nanoparticles. The results of the measurements have determined that Au–Ag multimetallic nanoplanets present large values of nonlinear optical parameters, such as the nonlinear refractive index that was determined in a negative magnitude of $n2 = 8 \times 10^{-10}$ cm^2/W [60]. Thus, it has been determined that, at intensities below the ablation threshold, Au–Ag nanoplanets behave as a saturable absorbent.

Otherwise, the nonlinear optical response of Au–Ag nanoplanets in silica has also been investigated. Optical measurements have defined that metallic Au–Ag nanoplanets immersed in silica influence the nonlinear absorption properties of the samples such that the nonlinear absorption coefficient determined in this way gives as a result $\beta = 1.1 \times 10^{-4}$ cm/W [61].

6.14   Nanostars Second- and Third-Order Optical Nonlinearities

Nanostars are a type of nanoparticle that consists of a nanoparticle with a spherical-based morphology with multiple arms. These types of nanoparticles resemble the shape of a star as shown in Fig. 6.7.

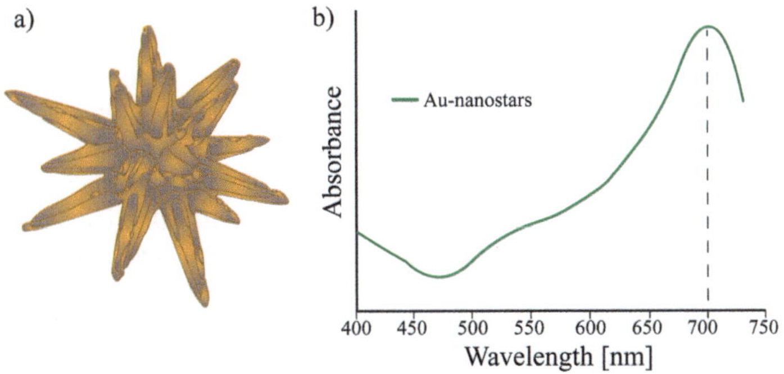

Fig. 6.7 (**a**) Geometrical representation of Au nanostars; (**b**) UV–VIS spectrum of Au nanostars

In nanostars, the response of second-order optical nonlinearities has a very high first-order optical hyperpolarizability. The former can be up to 130 times higher compared to other forms of gold nanoparticles. These improvements are attributed to the presence of sharp points on the surface such that the magnitude of the first hyperolarizability associated with gold nanostars is 78.6×10^{-25} esu [62].

Nanostars are mainly made up of metallic alloys, such as Au, Ag, Pt, Pd, and Rh. Moreover, the nanostars can also be composed of other elements such as silica, which can also be used to coat the nanoparticles. In optical measurements made to metallic Au nanostars with a size of 100 nm, it has been described that the nanostars have a multiple response to surface plasmon resonance. The localized surface plasmon resonances have sensitivity to the local dielectric environment due to the tips of the star-shaped structure. The aforementioned very strong dielectric sensitivity and near-infrared resonances define nanostars as very good substrates for sensors with surface plasmon resonance responses. The above are considered ideal characteristics for biological detection applications. The optical properties associated with Au nanostars are associated with the improvement of biological and photonic applications. Furthermore, nanostars provide multispectral signals, with the possibility of detecting three-dimensional orientation through the method of the multidirectional polarized scattering exploitation [63].

As stated above, it is established that nanostars have the utility of controlling the performance of applications that are based on the resonance response of localized surface plasmon resonance. That is why it is necessary to study the effect of the geometry of nanostars on the transmission effect and optical intensity of the near field. So, there are studies that demonstrate the dependence of the material of the nanostars on the transmittance spectra. The metals that showed the best responses for the optical response related to the morphology of the nanostars were the metallic nanoparticles of Au, Ag, and Cu. Without considering the geometric structure of the nanostars, the Au and Cu nanoparticles showed near-field improvements like those of Ag. In another way, the Al nanoparticles did not show such encouraging results in the face of the improved electric field in applications related to the visible part of the electromagnetic spectrum because it shows relatively weak electric fields [64].

So far, it has been associated that the responses of localized surface plasmon resonances related to metallic nanostructures are closely related to the geometry of the nanoparticles, so that this property is of great improvement in the optical nonlinear response.

The optical responses have been studied due to the geometry of some types of nanostars, which result in differences in nonlinearities due to the morphology of the nanoparticles such that the nonlinear refraction of nanostars with spherical structure turns out to be greater than that of nanostars with long branching geometry at a wavelength of 563 nm. However, at wavelengths between 700 nm in nanostars with long branching geometry, the nonlinear refraction results are higher. The result of the nonlinear refractive index of long branching nanostars at a wavelength of 706 nm, is $n_2 = 5.86 \times 10^{-4}$ cm^2/GW. Otherwise, the result of the refractive index of the nanostars has a magnitude of $n_2 = 3.92 \times 10^{-4}$ cm^2/GW. As can be seen, the nonlinear optical results of long ramification nanostars are up to 1.5 times greater than those of spherical nanostars [65].

The results of the optical nonlinearities of Au nanostars have shown a great advance in the synthesis methods of peculiarly complex nanoparticles. Besides, a panorama is opened in which noble metal nanoparticles promise improved optical properties in novel applications, as has happened with magnetic and semiconductor nanomaterials.

6.15 Nanopores Third-Order Optical Nonlinearities

Nanoporous are small holes about 1 nanometer in internal diameter. The synthesis method of nanopores consists of generating nanometric holes on solid nanostructures, such as silicon and graphene.

The synthesis of nanopores lies in generating the nanopores in a conductive medium; when applying a voltage, an electric current produced by the conduction of ions through the hole can be observed. Moreover, nanopores can be created through biological methods such as the use of proteins that have the function of perforating membranes.

The nonlinear optical response for nanopores layers of titanium dioxide has been studied through pulses in the ps regime and with a wavelength of 1064 nm. The result of third-order nonlinear optical susceptibility of the nanopores anatase layers exhibits a magnitude of $\chi^{(3)} = 2 \times 10^{-5}$ esu [66].

The results have shown significant differences between the nonlinearities studied for bulk anatase titanium dioxide nanoparticles. The differences in nonlinearities show that nanopores titanium dioxide membranes are up to six orders of magnitude greater. The magnitude of third-order nonlinear optical susceptibility can be explained through the resonant excitation response of the bandgap states of titanium dioxide. Thus, it can be predicted that the nonlinear optical response of anatase titanium dioxide could increase if additional defects are generated in the nanopore membranes of the material.

Besides, porous silicon has become a material of interest due to its remarkable nonlinear optical properties. Therefore, this material has malleable characteristics such as the control of the refractive index due to the process of synthesis of porous silicon.

Also, the nonlinear absorption and refraction coefficients of nanopores silicon have been observed. It is observed that the resulting ultrafast nonlinearities are attributed to the two-photon absorption mechanism. Also, improved nonlinear coefficients are observed for third-order optical nonlinearities in porous silicon. In addition, increases of more than twofold in third-order nonlinear optical nonlinearities have also been observed in crystalline silicon nanopores membranes [67].

6.16   Nanoholes Third-Order Optical Nonlinearities

In a different geometric case, the optical properties and the response of surface plasmon resonances in rectangular plasmonic nanoholes have been studied. The results show that the third harmonic nonlinear response increases significantly. Furthermore, the bandwidth of the harmonic that produces a potential light source in the ultraviolet range is widened [68].

On the other hand, structures with circular and triangular double nanoholes have also been investigated through the generation of the third harmonic, in which it has been established that the nonlinearities of double circular geometries are significantly higher than triangular ones. Also, the influence of the sizes of the nanoholes on the nonlinearities of the nanomaterials has been determined. The results have shown that in changes of increase in the size of the nanoholes in the range of 45 nm, the response of the nonlinearities of the nanomaterial was lower such that, in the case of nanoholes with a size of approximately 30 nm, the nonlinearities increase [69]. This shows that the size of nanoholes has a direct impact on the nonlinear optical response of nanomaterials.

The nonlinear optical properties of nanoholes made on gold and aluminum nanofilms have been studied. The third-order nonlinear optical susceptibility response due to third harmonic generation on nanofilms has been studied. Third-order nonlinear optical susceptibility at 1560 nm wavelength with 120 fs pulses has been shown to be very favorable. The third-order nonlinear optical susceptibility of simple circular Au nanoholes has been shown to be of magnitude $\chi^{(3)} = 2.15 \times 10^{-13}$ esu. Moreover, the third-order nonlinear optical susceptibility of simple circular Al nanoholes shows a value of $\chi^{(3)} = 4.31 \times 10^{-13}$ esu [69]. Moreover, the presence of a strong localized surface plasmon resonance of a single nanohole is demonstrated in the aluminum films, which results in a potential efficient nanolocated radiation source at the third harmonic frequency (Fig. 6.8).

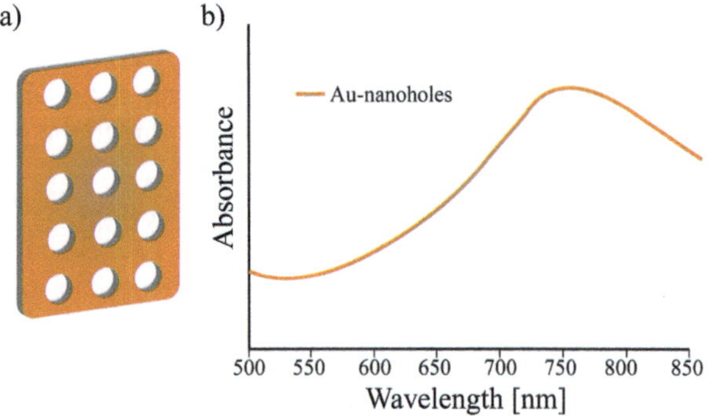

Fig. 6.8 (**a**) Geometrical representation of Au nanoholes; (**b**) UV–VIS spectrum of Au nanoholes

6.17   Highlights of Optical Nonlinearities in Nanocrystals and Complex Geometries

Semi-organic crystals can combine the most outstanding characteristics of organic and inorganic materials such that studies based on crystals as nonlinear optical materials are of importance for their potential applications in the field of laser technology and quantum data storage. An example of this is the use of crystals for the operation of high-energy laser systems. Some of the most used crystals are potassium dideuterium phosphate, lithium niobate, and barium titanate. The crystals have been used for the development of applications involving laser frequency doublers, electro-optical modulators operating in the infrared range, and phase conjugation. Furthermore, the saturable absorption exhibited by lithium niobate crystals can be used for ultrafast optical modulation in laser oscillators to generate ultrashort laser pulses. Besides, the nonlinearities exhibited by potassium thiourea chloride single crystals are attractive for applications related to night-vision sensors, optical limiters, and photonic devices. Similarly, high-speed data processing is essential for many applications in the field of computing and telecommunications. In addition, third-order nonlinear optical response is paramount for quantum computing applications related to switching. Optical architectures are very beneficial for high bit rate serial and parallel data stream processing systems. It is important to emphasize that the efficiency of optical data processing processes is closely dependent on the material used for the coupling of electrical and optical signals.

Organic materials and inorganic semiconductors are also good candidates for the development of nonlinear optical systems applications. Furthermore, organic materials are of great interest due to their low cost and ease of integration in device manufacturing. The plasmonic properties of the metallic nanoparticles allow increasing the emission and the control in the optical response due to the excitation of the electrons of the metallic structures. Furthermore, plasmonic behavior derived from decorations with metallic nanoparticles can function as an oscillator for frequency modulation. Also, gold nanoflowers have a high conductive capacity due to their size and morphology. Because of this, potential materials are considered in the design and application of energy storage devices. On the other hand, metallic nanostars are considered ideal for biological detection applications. This statement is because they have high dielectric sensitivity, as well as near-infrared resonances. This is what defines nanostars as ideal structures for the development of substrates sensitive to surface plasmon resonances.

References

1. Koukourakis, N., Abdelwahab, T., Li, M. Y., Höpfner, H., Lai, Y. W., Darakis, E., & Hofmann, M. R. (2011). Photorefractive two-wave mixing for image amplification in digital holography. *Optics Express, 19*(22), 22004–22023.
2. Briggs, I., Hou, S., Cui, C., & Fan, L. (2021). Simultaneous type-I and type-II phase matching for second-order nonlinearity in integrated lithium niobate waveguide. *Optics Express, 29*(16), 26183–26190.
3. Ding, Y., Osaka, A., Miura, Y., Toratani, H., & Matsuoka, Y. (1995). Second order optical nonlinearity of surface crystallized glass with lithium niobate. *Journal of Applied Physics, 77*(5), 2208–2210.
4. Wang, C., Zhang, M., Yu, M., Zhu, R., Hu, H., & Loncar, M. (2019). Monolithic lithium niobate photonic circuits for Kerr frequency comb generation and modulation. *Nature Communications, 10*(1), 1–6.
5. Joulaud, C., Mugnier, Y., Djanta, G., Dubled, M., Marty, J. C., Galez, C., & Le Dantec, R. (2013). Characterization of the nonlinear optical properties of nanocrystals by Hyper Rayleigh Scattering. *Journal of Nanobiotechnology, 11*(1), 1–9.
6. Gokila, G., Aarthi, R., & Raja, C. R. (2020). Elucidation of the properties of Lithium bis (2-methyllactato) borate monohydrate crystal for laser applications. *Journal of Materials Science: Materials in Electronics, 31*(9), 6956–6962.
7. Tumuluri, A., Bharati, M. S. S., Rao, S. V., & Raju, K. J. (2017). Structural, optical and femtosecond third-order nonlinear optical properties of LiNbO3 thin films. *Materials Research Bulletin, 94*, 342–351.
8. Karuna, S., Balu, A. R., Shyamala, D., Nagarethinam, V. S., & Delci, Z. (2018). Growth and characterization of third-order nonlinear optical lithium hydrogen maleate dihydrate single-crystal. *International Journal of Modern Physics B, 32*(31), 1850351.
9. Shanon, Z. S., Alnayli, R. S., & Tahir, K. J. (2016). Study of the second and third harmonics generation in lithium triborate single crystal. *International Journal of Science and Research (IJSR)*, 1614–1161.
10. Barbosa-Silva, R., Silva, J. F., Rocha, U., Jacinto, C., & de Araújo, C. B. (2019). Second-order nonlinearity of NaNbO3 nanocrystals with orthorhombic crystalline structure. *Journal of Luminescence, 211*, 121–126.
11. Poirier, G., Dussauze, M., Rodriguez, V., Adamietz, F., Karam, L., Cardinal, T., & Fargin, E. (2019). Second harmonic generation in sodium tantalum germanate glasses by thermal poling. *The Journal of Physical Chemistry C, 123*(43), 26528–26535.
12. Tong, A. S., Bondu, F., Senthil Murugan, G., Wilkinson, J. S., & Dussauze, M. (2019). Effect of sodium addition and thermal annealing on second-order optical nonlinearity in thermally poled amorphous Ta2O5 thin films. *Journal of Applied Physics, 125*(1), 015104.
13. Ravisankar, R., Jayaprakash, P., Eswaran, P., Mohanraj, K., Vinitha, G., & Pichumani, M. (2020). Synthesis, growth, optical and third-order nonlinear optical properties of glycine sodium nitrate single crystal for photonic device applications. *Journal of Materials Science: Materials in Electronics, 31*(20), 17320–17331.
14. Jagannath, G., Eraiah, B., Gaddam, A., Fernandes, H., Brazete, D., Jayanthi, K., & Allu, A. R. (2019). Structural and femtosecond third-order nonlinear optical properties of sodium borate oxide glasses: Effect of antimony. *The Journal of Physical Chemistry C, 123*(9), 5591–5602.
15. Packiya Raj, M., Ravi Kumar, S. M., Sivavishnu, D., Kubendiran, T., Anbarasi, A., & Allen Moses, S. E. (2018). Synthesis, growth and optical, mechanical, electrical and surface properties of an inorganic new nonlinear optical crystal: Sodium cadmium tetra chloride (SCTC). *Crystal Research and Technology, 53*(2), 1700271.
16. Sundari, A., & Manikandan, S. (2018). Third-order nonlinear optical properties and optical limiting behaviour of sodium penta borate. *Journal of Materials Science: Materials in Electronics, 29*(1), 558–567.

17. Pernice, P., Paleari, A., Ferraris, M., Fokine, M., Fanelli, E., Lorenzi, R., & Aronne, A. (2009). Electric field induced structural modification and second order optical nonlinearity in potassium niobium silicate glass. *Journal of Non-Crystalline Solids, 355*(52–54), 2578–2582.

18. Tanaka, H., Yamamoto, M., Takahashi, Y., Benino, Y., Fujiwara, T., & Komatsu, T. (2003). Crystalline phases and second harmonic intensities in potassium niobium silicate crystallized glasses. *Optical Materials, 22*(1), 71–79.

19. Sigaev, V. N., Stefanovich, S. Y., Champagnon, B., Gregora, I., Pernice, P., Aronne, A., & Dewhurst, C. (2002). Amorphous nanostructuring in potassium niobium silicate glasses by SANS and SHG: A new mechanism for second-order optical nonlinearity of glasses. *Journal of Non-Crystalline Solids, 306*(3), 238–248.

20. Azhar, S. M., Rabbani, G., Shirsat, M. D., Hussaini, S. S., Baig, M. I., Ghramh, H. A., & Anis, M. (2018). Luminescence, laser induced nonlinear optical and surface microscopic studies of potassium thiourea chloride crystal. *Optik, 165*, 259–265.

21. Dhatchaiyini, M. K., Rajasekar, G., & Bhaskaran, A. (2019). Synthesis, growth, optical, photoconductivity and specific heat properties of potassium borodicitrate (KBDC) single crystal. *Journal of Materials Science, 54*(13), 9362–9371.

22. Balakrishnan, C., Manonmani, M., Vinitha, G., Meenakshisundaram, S. P., & Sockalingam, R. M. (2020). Effect of organic additive on the growth and characterization of potassium hydrogen phthalate crystals: Third-order nonlinear optical properties. *Materials Today: Proceedings, 34*, 425–429.

23. Raja, M. D., Pari, S., Kanagan, G., Kumar, G. S., Christuraj, P., & Rajkumar, D. (2022). Optical, spectral and NLO properties of L-Leucine potassium chloride crystals. *Materials Today: Proceedings, 50*(7), 2687–2690.

24. Uthayakumar, M., Saraswathi, V., Pasupathi, G., Manimekalai, R., & Shinde, V. (2020). Structural, third-order nonlinear optical properties and optical limiting studies of novel zinc potassium aluminum sulfate nonadecahydrate single crystal. *Journal of Materials Science: Materials in Electronics, 31*(24), 22522–22533.

25. Gerstenberger, D. C., Trautmann, T. M., & Bowers, M. S. (2003). Noncritically phase-matched second-harmonic generation in cesium lithium borate. *Optics Letters, 28*(14), 1242–1244.

26. Sifi, A., Klein, R. S., Maillard, A., Kugel, G. E., Péter, A., & Polgár, K. (2003). Absolute nonlinear optical coefficients measurements of CsLiB6O10 single crystals by second harmonic generation. *Optical Materials, 24*(1–2), 431–435.

27. Liu, S., Chen, G., Huang, Y., Lin, S., Zhang, Y., He, M., & Liang, X. (2017). Tunable fluorescence and optical nonlinearities of all inorganic colloidal cesium lead halide perovskite nanocrystals. *Journal of Alloys and Compounds, 724*, 889–896.

28. Ahamed, S. R., Srinivasan, P., Balaji, J., Raj, S. G., & Mohan, S. (2017). Structural, spectral, thermal, micro hardness, dielectric and etching studies of third order nonlinear optical material Cesium Sulfamate. *Journal of Alloys and Compounds, 701*, 822–827.

29. Sudha, S., Kumari, C. R. T., Nageshwari, M., Ramesh, P., Vinitha, G., Caroline, M. L., & Manikandan, A. (2020). Crystal growth, optical, spectroscopic studies, PL behaviour and Hirshfield surface analysis of a third-order nonlinear optical Cesium Hydrogen Oxalate Dihydrate (CHOD) single crystal. *Journal of Materials Science: Materials in Electronics, 31*(18), 15028–15037.

30. Hegde, T. A., Dutta, A., Jauhar, R. M., Karuppasamy, P., Pandian, M. S., Abith, M., & Vinitha, G. (2020). Physicochemical properties of cesium tetroxalate dihydrate single crystal: An efficient material for nonlinear optical applications. *Optical Materials, 107*, 110033.

31. Balasubramanian, D., Sankar, R., Shankar, V. S., Murugakoothan, P., Arulmozhichelvan, P., & Jayavel, R. (2008). Growth and characterization of semiorganic nonlinear optical rubidium bis-dl-malato borate single crystals. *Materials Chemistry and Physics, 107*(1), 57–60.

32. Debnath, R., Beda, S. K., Hada, D. S., & Saha, A. (2017). Electro-optically induced nonlinear phase shift in RTP crystal by cascaded second-order nonlinearity. In *2017 Conference on Lasers and Electro-Optics Pacific Rim (CLEO-PR)* (p. s1789). Optical Society of America.

33. Terashima, K., Kim, S. H., & Yoko, T. (1995). Nonlinear optical properties of B_2O_3-based glasses: M_2O-B_2O_3 (M= Li, Na, K, Rb, Cs, and Ag) binary borate glasses. *Journal of the American Ceramic Society, 78*(6), 1601–1605.
34. Sukumar, M., Babu, R. R., & Ramamurthi, K. (2016). Linear and third-order nonlinear optical properties of $LiRbB_4O_7$ single crystal. *Solid State Sciences, 51*, 8–12.
35. Li, Y. Q., & Xiao, M. (1996). Enhancement of nondegenerate four-wave mixing based on electromagnetically induced transparency in rubidium atoms. *Optics Letters, 21*(14), 1064–1066.
36. Celebrano, M., Biagioni, P., Finazzi, M., Duo, L., Zavelani-Rossi, M., Polli, D., & Cerullo, G. (2008). Near-field second-harmonic generation from gold nanoellipsoids. *Physica Status Solidi c, 5*(8), 2657–2661.
37. Ulriksen, H. U., & Pedersen, K. (2016). Field enhancement at silicon surfaces by gold ellipsoids probed by optical second-harmonic generation spectroscopy. *Journal of Applied Physics, 120*(23), 235307.
38. Boltaev, G. S., Ganeev, R. A., Krishnendu, P. S., Zhang, K., & Guo, C. (2019). Nonlinear optical characterization of copper oxide nanoellipsoids. *Scientific Reports, 9*(1), 1–8.
39. Mohan, S., Lange, J., Graener, H., & Seifert, G. (2012). Surface plasmon assisted optical nonlinearities of uniformly oriented metal nano-ellipsoids in glass. *Optics Express, 20*(27), 28655–28663.
40. Sun, X., Chen, X., Liu, X., & Qu, S. (2011). Optical properties of poly (methyl methacrylate)–titania nanostructure thin films containing ellipsoid-shaped titania nanoparticles from ex-situ sol-gel method at low growth temperature. *Applied Physics B, 103*(2), 391–398.
41. Chen, F., Dai, S., Xu, T., Shen, X., Lin, C., Nie, Q., & Heo, J. (2011). Surface-plasmon enhanced ultrafast third-order optical nonlinearities in ellipsoidal gold nanoparticles embedded bismuthate glasses. *Chemical Physics Letters, 514*(1–3), 79–82.
42. Das, K., Uppal, A., Saini, R. K., Varshney, G. K., Mondal, P., & Gupta, P. K. (2014). Hyper-Rayleigh scattering from gold nanoparticles: Effect of size and shape. *Spectrochimica Acta Part A: Molecular and Biomolecular Spectroscopy, 128*, 398–402.
43. Hubert, C., Billot, L., Adam, P. M., Bachelot, R., Royer, P., Grand, J., & Fort, A. (2007). Role of surface plasmon in second harmonic generation from gold nanorods. *Applied Physics Letters, 90*(18), 181105.
44. Bernasconi, G. D., Butet, J., & Martin, O. J. (2018). Dynamics of second-harmonic generation in a plasmonic silver nanorod. *ACS Photonics, 5*(8), 3246–3254.
45. Wang, X., Yao, L., Li, S., & Cai, M. (2020). Extraordinarily large third-order optical nonlinearity in Au nanorods under nanowatt laser excitation. *The Journal of Physical Chemistry C, 124*(12), 6838–6844.
46. Sato, R., Henzie, J., Rong, H., Naito, M., & Takeda, Y. (2019). Enhancement of the complex third-order nonlinear optical susceptibility in Au nanorods. *Optics Express, 27*(14), 19168–19176.
47. Dai, H., Zhang, L., Wang, Z., Wang, X., Zhang, J., Gong, H., & Han, Y. (2017). Linear and nonlinear optical properties of silver-coated gold nanorods. *The Journal of Physical Chemistry C, 121*(22), 12358–12364.
48. Zhou, Z. J., Li, X. P., Ma, F., Liu, Z. B., Li, Z. R., Huang, X. R., & Sun, C. C. (2011). Exceptionally large second-order nonlinear optical response in donor–graphene nanoribbon–acceptor systems. *Chemistry-A European Journal, 17*(8), 2414.
49. Gui, Z., Wang, X., Liu, J., Yan, S., Ding, Y., Wang, Z., & Hu, Y. (2006). Chemical growth of ZnO nanorod arrays on textured nanoparticle nanoribbons and its second-harmonic generation performance. *Journal of Solid State Chemistry, 179*(7), 1984–1989.
50. Attaccalite, C., Cannuccia, E., & Grüning, M. (2017). Excitonic effects in third-harmonic generation: The case of carbon nanotubes and nanoribbons. *Physical Review B, 95*(12), 125403.
51. Karimi, F., Davoody, A. H., & Knezevic, I. (2018). Nonlinear optical response in graphene nanoribbons: The critical role of electron scattering. *Physical Review B, 97*(24), 245403.

52. Li, J., Chen, W., Liu, J., Sun, W., Li, Z., & Li, Y. (2021). Effects of the nanowire length on large second-order nonlinear optical responses: A theoretical investigation of the thinnest doped beryllium nanowires with IR and UV working wavebands. *Dalton Transactions, 50*(13), 4613–4622.
53. Yuan, C., Li, X., Semin, S., Feng, Y., Rasing, T., & Xu, J. (2018). Chiral lead halide perovskite nanowires for second-order nonlinear optics. *Nano Letters, 18*(9), 5411–5417.
54. Wang, K., Zhou, J., Yuan, L., Tao, Y., Chen, J., Lu, P., & Wang, Z. L. (2012). Anisotropic third-order optical nonlinearity of a single ZnO micro/nanowire. *Nano Letters, 12*(2), 833–838.
55. Han, Y. P., Ye, H. A., Wu, W. Z., & Shi, G. (2008). Fabrication of Ag and Cu nanowires by a solid-state ionic method and investigation of their third-order nonlinear optical properties. *Materials Letters, 62*(17–18), 2806–2809.
56. Parida, M. R., Vijayan, C., Rout, C. S., Sandeep, C. S., Philip, R., & Deshmukh, P. C. (2011). Room temperature ferromagnetism and optical limiting in V2O5 nanoflowers synthesized by a novel method. *The Journal of Physical Chemistry C, 115*(1), 112–117.
57. Wei, R., Zhang, H., Tian, X., Qiao, T., Hu, Z., Chen, Z., & Qiu, J. (2016). MoS$_2$ nanoflowers as high performance saturable absorbers for an all-fiber passively Q-switched erbium-doped fiber laser. *Nanoscale, 8*(14), 7704–7710.
58. Feng, J., Li, X., Zhu, G., & Wang, Q. J. (2020). Emerging high-performance SnS/CdS nanoflower heterojunction for ultrafast photonics. *ACS Applied Materials & Interfaces, 12*(38), 43098–43105.
59. Pellegrini, G., Bello, V., Mattei, G., & Mazzoldi, P. (2007). Local-field enhancement and plasmon tuning in bimetallic nanoplanets. *Optics Express, 15*(16), 10097–10102.
60. Cesca, T., Calvelli, P., Battaglin, G., Mazzoldi, P., & Mattei, G. (2012). Local-field enhancement effect on the nonlinear optical response of gold-silver nanoplanets. *Optics Express, 20*(4), 4537–4547.
61. Cesca, T., Pellegrini, G., Bello, V., Scian, C., Mazzoldi, P., Calvelli, P., & Mattei, G. (2010). Nonlinear optical properties of Au–Ag nanoplanets made by ion beam processing of bimetallic nanoclusters in silica. *Nuclear Instruments and Methods in Physics Research Section B: Beam Interactions with Materials and Atoms, 268*(19), 3227–3230.
62. Nehl, C. L., Liao, H., & Hafner, J. H. (2006). Optical properties of star-shaped gold nanoparticles. *Nano Letters, 6*(4), 683–688.
63. Zhu, S., & Cortie, M. (2015). Optical properties of arrays of five-pointed nanostars. In *Micro+ nano materials, devices, and systems* (Vol. 9668, p. 96683R). International Society for Optics and Photonics.
64. Liu, X. L., Wang, J. H., Liang, S., Yang, D. J., Nan, F., Ding, S. J., & Wang, Q. Q. (2014). Tuning plasmon resonance of gold nanostars for enhancements of nonlinear optical response and Raman scattering. *The Journal of Physical Chemistry C, 118*(18), 9659–9664.
65. Gayvoronsky, V., Galas, A., Shepelyavyy, E., Dittrich, T., Timoshenko, V. Y., Nepijko, S. A., & Koch, F. (2005). Giant nonlinear optical response of nanoporous anatase layers. *Applied Physics B, 80*(1), 97–100.
66. Suess, R. J., & Murphy, T. E. (2012). Third-order optical nonlinearity in bulk nanoporous silicon at telecom wavelengths. In *CLEO: Applications and technology* (p. JW4A-43). Optical Society of America.
67. Shi, L., Andrade, J. R., Yi, J., Marinskas, M., Reinhardt, C., Almeida, E., & Kovacev, M. (2019). Nanoscale broadband deep-ultraviolet light source from plasmonic nanoholes. *ACS Photonics, 6*(4), 858–863.
68. Yi, T., Su, W., & Geng, Z. (2019). Third-harmonic generation from double nanohole aperture in gold film. *IEEE Photonics Technology Letters, 31*(24), 1936–1939.
69. Konstantinova, T. V., Melentev, P. N., Afanasev, A. E., Kuzin, A. A., Starikov, P. A., Baturin, A. S., & Balykin, V. I. (2013). A nanohole in a thin metal film as an efficient nonlinear optical element. *Journal of Experimental and Theoretical Physics, 117*(1), 21–31.

Chapter 7
Applications, Conclusions, and Perspectives

7.1 Second-Order Nonlinear Optical Applications

The understanding of the interactions between the physicochemical structures and the molecular nonlinearities is of relevant importance because through this it is possible to optimize the nonlinear optical properties of organic materials. As discussed before, second-order optical nonlinearities are mainly associated with materials with non-centrosymmetric geometric structures. The above also denotes that second-order nonlinearities result from geometries in which the ease of polarization in one direction is different from that of the opposite direction. To exhibit second-order optical nonlinearities, in addition to lacking an inversion center of symmetry, the materials must satisfy the standard phase coincidence. Some of the materials that exhibit significant optical nonlinearities are birefringent crystals that allow dispersion compensation. Moreover, electrical asymmetric organic molecules that have high first-order molecular hyperpolarizability are associated with second-order nonlinear optical effects.

On the other hand, for there to be a high second-order nonlinear response, the materials must be non-centrosymmetric. Besides, it is necessary that there be a difference between the dipolar moments of the fundamental state and that in the excited state it must be as large as possible. Furthermore, the crystalline packing must be asymmetric, such that the chiral molecules are usually optimal.

It has been determined that there are important factors for the application of the material in second-order nonlinear optical applications. These factors are the efficiency, which is related to the factors that affect the effectiveness of the second harmonic generation, the threshold that refers to the intensity of radiation that material supports, and the optical window that refers to the region of the UV–VIS spectrum where there is no absorption. It has also been established that there are two main types of optimal materials for the generation of second-order nonlinearities: mineral oxides and ferroelectric single crystals. Mineral oxides have advantages

© The Author(s), under exclusive license to Springer Nature Switzerland AG 2022
C. Torres-Torres, G. García-Beltrán, *Optical Nonlinearities in Nanostructured Systems*,
Springer Tracts in Modern Physics 287, https://doi.org/10.1007/978-3-031-10824-2_7

such as their great transparency in the UV–Vis spectrum and have a very high threshold of optical damage; however, they have a low coefficient of optical hyperpolarizability. Some examples of mineral oxides are potassium thiophosphate and lithium niobate. On the other hand, organic materials have a wide range of structures; in addition, organic compounds are composed of a donor and receptor of electrons joined through a covalent bond through a delocalized π-electron system.

However, as the devices that use the second harmonic generation nonchange from the macroscopic to the nanoscopic level, the symmetry considerations change. So, individual molecule symmetries must be considered rather than crystallographic inversion symmetry. This also defines a wide range of structures that can be analyzed for possible second harmonic generation applications in nonlinear optical nanodevices. In addition, there are methods for molecules that are intrinsically centrosymmetric to have utility in applications related to second-order optical phenomena. Just as the incorporation of a chiral center or the addition of a simple hydrogen bond can break the centrosymmetry of point groups of a molecule, in addition to the application of electric fields, turning the molecules into a viable system for nonlinear optical applications.

On the other hand, quadratic polarization causes singular effects, necessary to generate second-order nonlinear optical phenomena that consist of mixing wave phenomena, based on the generation of sum and difference frequencies. These wave-mixing phenomena are important for the generation of many applications, such as tunable sources, parametric amplifiers, and oscillators. Also, the presence of radiation at the frequency $\omega 2$ or $\omega 3$ can stimulate the emission of additional photons at these frequencies such that optical parametric oscillators based on this effect are used in tunable infrared source applications.

There are numerous applications derived from second-order optical nonlinearities, biological applications being among the main ones. The images derived from the second harmonic generation can be interleaved with fluorescence microscopy techniques. This is to obtain information on biological structures since multiphoton fluorescence microscopy requires an environment without a center of symmetry, and it is very sensitive to any contrast modification in any interfacial region that produces a signal derived from the second harmonic generation. Thus, the images derived from the second harmonic generation show improvements due to the chirality of the molecules. For example, in the particular case of cells, they have asymmetric membranes for optical beams since their size is greater than the optical wavelength.

On the other hand, the electronic properties of nanostructures, in combination with strong mechanical and thermal properties, are of great interest for the design of future electronic and optical applications. In addition, nonlinear optical techniques signal the response of the system to multiple interactions with the electromagnetic field. In addition, optical nonlinearity measurement techniques are especially interesting because nanomaterials can have remarkably large nonlinear optical responses and through them provide complementary information. In the particular case of second-order processes, they are highly sensitive to symmetry, and non-centrosymmetric materials intervene in the electric dipole approximation of the light–matter interaction. In addition, chirality is a structural property that breaks structural centrosymmetry and allows second-order optical processes. Thus, carbon nanotubes have been proposed as an optimal chiral nanostructure for second-order nonlinear optical processes. In addition, this property is important for the development of applications associated with molecular recognition. In addition, organic nanomaterials have attracted interest due to the polarizability of electrons in addition to a fast response. However, they do not have good thermal stability, unlike carbon allotropes, which have good electron conjugation and thermal stability. Thus, there is a wide field of application for nanomaterials with the second-order nonlinear optical response.

On the other hand, the second harmonic generation technique uses nonlinear electronic polarization of molecules through laser pulse excitation such that coherent radiation of a different color from that of the incident laser light beam is generated, which provides detection by spectral separation. Such applications range from frequency conversion, data, and image storage, as well as applications in the field of optical and biological sensors and switches [1]. In particular, it is very useful in applications related to biological systems, such as collagen distribution and membrane detection [2].

On the other hand, crystallized metal chalcogenides, also called tetrahedral chalcogenides, have special structural characteristics and adjustable optical bandgaps. This makes them excellent candidates for the nonlinear optical response, in addition to photocatalytic, photovoltaic, and thermoelectric applications. This is mainly due to the valence and conduction bands derived from the Cu-3d and Sb5s orbitals, in addition to the S-3p orbital that contributes to both bands. Furthermore, the high S–S and Sb–Sb transformation contributions lead to the response of high absorption coefficients. This indicates that crystallized metal chalcogenides are excellent for photovoltaic applications [3].

On the other hand, semiconductor quantum wells have exhibited strong second-order $\chi^{(2)}$ optical nonlinearities; however, they possess limited confinement potential. This represents a challenge in frequency applications in the near-infrared frequency range. Thus, metal/dielectric-based nanoheterostructures possess a high sensitivity at near-infrared frequencies. As an added advantage, the heterostructure in combination with an electric field enhancement results in a nanostructured surface, which in turn exhibits energy efficiency in the second harmonic generation by several orders of magnitude. Also, heterostructures enable efficient optical wave mixing in ultra-compact nonlinear optical devices. This application is achieved through plasmonic enhancements of the electric field whose wavelength is below the wavelength of light.

Such enhancement of the electric field through plasmonic properties can be combined with the high second-order nonlinearities $\chi^{(2)}$ for the second harmonic generation of a surface at the visible frequency by constructing a nanostructured surface with quantum dots. Equally, through these nanometric surfaces, a transverse electric field polarization conversion phenomenon is possible, which allows the normal excitation of the incident beam to expand the usability of quantum dots with extremely high nonlinear responses [4].

On the other hand, it has been established that nonresonant excitations through gallium selenide waveguides contribute to the enhancement of second-order nonlinear optical processes associated with second harmonic generation and sum-frequency generation such that the improvement of the optical process derived from the integration of the gallium selenide is established in up to four orders of magnitude. This establishes that the efficiency of the frequency conversion is also improved with the perfect coating of 2D nanostructured materials such that a strong and tunable light–matter interaction is facilitated. In addition, the applications related to frequency converters are integrated with the telecommunications infrastructure. Thus, its possible applications in fiber laser sources with frequency conversion and optical signal processing are highlighted [5].

The second harmonic generation technique has been used for the observation of biological samples. This technique has been very useful in the selective observation of membranes, in addition to the sensitive detection of membrane damage. On the other hand, it has been established that the intensity of the second harmonic generation depends on the molecular alignment. Further, through the second harmonic generation, it is possible to observe the active components of biological tissues, in addition to evaluating the damage to the membrane [6]. However, the application of nonlinear optics to biology is not limited to imaging.

Moreover, the second harmonic generation is a highly sensitive technique with the ability to detect proteins, peptides, and small molecule adsorption on a surface. This is because, since there are strict rules about symmetry in second-order nonlinear optical processes, extreme surface specificity is established, eliminating contributions from diffusion or molecular binding of the bulk solution such that the characteristics associated with the second harmonic generation related to versatility, high sensitivity, and surface specificity in the second harmonic generation technique are great advantages for the detection of biosensors and biochemistry at interfaces [7].

On the other hand, the second harmonic generation has been used in the visualization of sections of healthy and cancerous human tissue. The above is to diagnose different types of cancer through imaging. These images exhibit alterations in tissue architecture as well as morphological changes at the subcellular level. This is established through the assignment of virtual colors to mimic hematoxylin, eosin, and saffron staining. Thus, it is possible to use the second harmonic generation technique as an additional tool for making medical decisions related to surgery [8].

In addition, through the second harmonic generation, it is possible to evaluate the structure of collagen. This is to evaluate the progression of cancer because the extracellular matrix associated with collagen changes [9].

7.2 Third-Order Nonlinear Optical Applications

Third-order nonlinear optical processes result from the third-order contribution to nonlinear polarization. One of the most important nonlinear optical processes is the third harmonic generation, which describes a process in which a wave with frequency 3ω is drifted through an incident field with frequency ω. This process results from the dissolution of three incident photons with frequency ω and from the simultaneous creation of one photon with frequency 3ω. It has also been mentioned that one of the experimental ways of generating a third-order nonlinear phenomenon is through a mixing process of four optical waves. In this process, there are three input optical waves with frequencies $\omega1$, $\omega2$, and $\omega3$, which interact with the material and produce a signal with frequency $\omega4$. Also, the wave output derived from the four optical wave mixing processes can be improved due to the vibratory or electronic properties of the materials. Also, when the incident light field satisfies the electron transition resonance condition of the material, an enhanced response is generated from third-order nonlinear response processes.

On the other hand, it is also established that third-order nonlinear optical susceptibility is mathematically represented through a tensor such that the number of independent components of the tensor can be reduced due to the intrinsic symmetry of the material, and therefore, it is associated that third-order nonlinearities are closely related to the geometric structure of materials.

On the other hand, the materials that exhibit third-order nonlinear optical responses are established, among which are insulating crystals, semiconductors, organic materials, and metallic and carbon nanostructures.

Otherwise, the third-order nonlinear optical susceptibilities exhibited by semiconductors depend on the wavelength, in addition to the energy of the incident photon and the bandgap of the semiconductor. This is because when the photon energy is less than the bandgap energy, there is little difference between semiconductors and insulating crystals. Excitation resonance is the main nonlinear optical mechanism in semiconductors when the photon energy is close to the bandgap energy. In the case where the photon energy is greater than the bandgap, the nonlinear response results from the excitation of the electron from the valence band to the conduction band.

Furthermore, the importance of organic materials is reflected due to the fact that large third-order nonlinear optical effects can be obtained through composite or hybrid materials. An example of this is the improvement of third-order optical susceptibility due to the embedding of metallic or semiconductor nanoparticles in glass substrates.

On the other hand, the materials that exhibit third-order nonlinear optical responses are established, among which are insulating crystals, semiconductors, organic materials, and metallic and carbon nanostructures.

The nonlinear third-order optical susceptibilities exhibited by semiconductors depend on the wavelength, in addition to the energy of the incident photon and the forbidden band of the semiconductor, which is because when the photon energy is less than the bandgap energy, there is almost no difference between semiconductors and insulating crystals. When the photon energy is close to the bandgap energy, it is considered as the main nonlinear optical mechanism in semiconductors. Furthermore, the importance of organic materials is reflected because large third-order nonlinear optical effects can be obtained through composite or hybrid materials. An example of this is the improvement of third-order optical susceptibility due to the embedding of metallic or semiconductor nanoparticles in glass substrates.

Derived from the above, great progress has been made in the application of third-order nonlinear optics to integrated optical devices, such that complementary optical devices to linear devices have been developed, in addition to new devices derived from strong nonlinearities. On the other hand, it has been observed that second-order and even orders of nonlinear optical effects are exhibited only through non-centimeter materials. However, third-order nonlinear optical effects can occur in centrosymmetric or non-centrosymmetric materials. The most widely used third-order nonlinear optical effects include third harmonic generation, four-wave mixing, and the optical Kerr effect. These third-order nonlinear optical effects are widely used in applications related to photon generation, frequency manipulation, and images.

On the other hand, the possibility of optical data storage through the third harmonic generation technique in silver zinc phosphate glass has been indicated. Data storage within the glass occurs through femtosecond laser irradiation below the refractive index modification threshold. In the same way, a nonlinear resonant interface is formed due to the creation of stable silver groups, derived from the accumulation effect. Besides, the signal derived from the third harmonic generation technique is coherent, and, in addition, an intense, directional, and less divergent beam is generated with a high signal/noise ratio [10].

Also, it has been established that for the design of applications related to optical switching, materials with large third-order nonlinearities and low optical losses are necessary. That is why, for the design of these applications, it is necessary to consider high third-order polarizability in addition to a molecular length control in order to obtain a favorable absorption of one and two photons [11].

Similarly, broadband semiconductors are considered excellent candidates for the design of various nonlinear optical and optoelectronic applications, such as optical filters, and optical and photovoltaic sensors. In addition to the fact that broadband semiconductors have exhibited excellent nonlinear optical and electrical properties, they have an adaptive advantage when faced with doping. So, modulation in bandgap and photoluminescence properties through doping of semiconductors with sodium and potassium has been exhibited. Such enhancements and control over optical and emission properties are of importance to produce low-power, high-performance optical devices [12].

The interaction of conduction electrons with electric fields gives rise to the optical properties of metallic nanostructures. Likewise, the localized surface plasmon resonance depends on the size and shape of the nanostructures, as well as the medium that surrounds them. In addition, nanostructures with complex noble metal geometries present multiple surface plasmon resonances due to their profiled tips, which contribute to enhancing the electric field due to lightning rod effects, causing a very high enhancement of the metallic optics. In addition, the response of third-order nonlinearities can be given as a function of effective absorption, nonlinear scattering, and nonlinear refractive index changes derived from light absorption mediated by metal nanostructures. From this, innovative applications have been found in a wide range of fields, including sensing, biology, electronics, and catalysis. Some other applications include biomedical diagnostics and therapy and biomedical laser-induced photothermal processes. Therefore, it is necessary to understand the mechanisms of energy absorption in metallic spherical nanostructures, in addition to the nonlinear characteristics of the materials excited by a laser wavelength close to the resonance absorption peak [13].

On the other hand, there is great interest in developing organic compounds with third-order nonlinear optical properties due to their applications in data storage, optical communications, all-optical photonic switching, and optical limiting devices. This is because third-order nonlinear organic materials with π-conjugated structures exhibit a strong nonlinear response due to intermolecular donor–acceptor interactions and delocalized π-electron systems. Thus, the real and imaginary part of the third-order nonlinear optical susceptibility is essential in the development of optical switching applications [14]. While the effects of optical power limitation are of importance in applications for eye protection and sensitive optical devices against high power laser pulses [15].

On the other hand, the third-order generation nonlinear optical process allows a variety of applications related to optical image processing. Such applications are achieved by taking advantage of the ultrafast nature of the third-order nonlinear process in order to generate a response time limited by the duration of the ultrafast pulses used to implement them. Besides, the third harmonic generation can be implemented for optical processing applications with large spectral and angular bandwidths. Otherwise, isotropic polymeric compounds are useful for optical processing because they have a strong structure and can form different structures with various thicknesses. Also, through the third harmonic generation technique, signals can be easily filtered both spatially and spectrally. Thus, frequency conversion processes are possible, which facilitates the development of applications compatible with silicon-based electronics when the wavelength of the signal beams is in the near-infrared range. In the same way, the large nonlinearities expected in branched molecules, dendrimers, and polymers based on octupolar units are attractive to the branch of developing centrosymmetric materials that can contribute to applications associated with third-order optical nonlinearities [16].

Furthermore, the third harmonic generation technique is easy to implement in multiphoton imaging platforms, which allows its application in fluorescence imaging and multiphoton infrared microscopy. In addition, some other fields of biological applications include neuroscience, biomaterials, vascular biology, cell flow kinetics, immunobiology, and revealing tissue structures that guide and determine the positioning of cancer and immune cells [17].

7.3   High-Order Nonlinear Optical Applications

There are high-order nonlinear optical processes that occur when the intensity of the laser field is comparable to the strength of the atomic bond. One of the optical processes for the reproduction of high-order nonlinearities is the high-harmonic generation. The high-harmonic generation is characterized by its usefulness in time-resolved photon and electron spectroscopy, as well as the generation of attosecond pulses. Also, high-harmonic generation is usually observed through rare gas atoms; however, this process is obtained with relatively low efficiency. On the other hand, bulk solids exhibit higher efficiency against high-harmonic generation than rare gases since bulk solids have a structure with a higher density of atoms. However, the efficiency of high-harmonic generation in solids is currently limited due to the strong absorption of the generated signals whose photon energy is greater than the forbidden band of nonlinear materials. Thus, the emission of a high fluence of the laser beam is necessary.

On the other hand, the improvement of the local field induced by the plasmonic resonance has been used to increase the response in the high-harmonic generation such that, with field enhancement supported by surface plasmon polaritons, high-order harmonics are generated.

On the other hand, the chiral properties of some molecules can benefit the response in the high-harmonic generation. Furthermore, the polarization state of attosecond pulses can be controlled by adjusting the time scale-weighted chirality of the laser beam. Thus, through the application of the high-harmonic generation technique, it is possible to develop high-order nonlinear applications in the field of chiral spectroscopy. In addition, it is also applicable in the design of applications related to the ionization of atoms and molecules in strong fields, production of electron vortex controllers, currents of rotational electrons, and polarized spin electrons.

Through high-order optical nonlinearities, it is possible to establish an effect known as automatic phase modulation that gives rise to an observed intensity pattern in the far field. This is due to the intensity-dependent refractive index and the interaction of the beam with the medium that produces changes in the phase of the beam.

On the other hand, spatial modulation instability is a phenomenon that manifests itself through the interaction between nonlinearities and diffraction effects, which in turn are enhanced by the growth of the optical field amplitude and noise. In addition, the spatial modulation instability is affected by high-order dispersive effects, saturation nonlinearity, nonlocal nonlinearity, and coherence properties of optical beams [18].

High-order harmonic generation is an optical process involving the emission of a burst of coherent radiation, collinear to the driving beam, with a characteristic comb-shaped spectrum of odd harmonics of the fundamental laser field, ranging from the ultraviolet of vacuum up soft X-rays. Through the application of the high-harmonic generation technique, it is used to generate attosecond pulses. Such that, through it, it is possible to study ultrafast dynamics in atoms, molecules, and condensed matter with temporal resolutions on the attosecond time scale. Evenly, the generation of high-order harmonics in microfluidic devices is possible [19]. Thus, the third harmonic generation technique is used to detect cell–tissue interfaces in a noninvasive manner by filling the resolution gap between images using high-resolution ultrasound, scanning electron microscopy to map the organization and tissue function in 3D [20].

Also, from this technique, microstructures are developed in transparent materials by focusing on attosecond laser pulses such that, in the focal region, nonlinear optical processes occur, such as multiphoton ionization, which implies the modification of the structural and chemical properties of the material. In this way, the phenomena of laser-induced nanostructuring are established, which in turn result in the chemical etching of the irradiated regions [21].

Likewise, femtosecond laser irradiation allows the realization of complex hollow microstructures, which also allows flexibility characteristics to be attributed to the manufacture of chips. These devices are used in applications related to the transport and control of liquids and gas flows [22].

The high-harmonic generation is based on tunneling ionization and also radiative recombination during a single optical cycle of the excitation pulse. Through it, optical phenomena caused by the interaction of light waves with electrons on the attosecond time scale are manifested. Then, through the high-harmonic generation, it is possible to observe attosecond quantum dynamics in atoms and molecules [23].

It has been established that the intensity of a specific harmonic order could be increased by one or two orders of magnitude through higher-order optical nonlinearities. This is observed when the wavelength of a harmonic is very close to an autoionizing resonance of the nonlinear medium. Also, these experiments have only shown that the intensity of the high-harmonic generation of the ellipticity of the conductive laser is established through phase coincidence conditions similar to those of other conventional harmonics [24].

7.4 Conclusions

This work evaluates the nonlinear optical interactions of the second-, third-, and high-order in nanostructures of diverse chemical and geometric structures. The theoretical–experimental aspects of each of the orders studied were described, in addition to the aspects of synthesis and characterization of various nanostructures.

Furthermore, the theoretical models that describe the influence of nonlinearities through effects such as quantum confinement, plasmonic phenomena, Fano resonances, the Purcell effect, and some physical mechanisms that include molecular orientation, electrostriction, and thermal effects were explored. An extensive analysis of existing and potential nanostructures and applications in the field of optical nonlinearities based on their specific characteristics for each order was also analyzed. The evaluations were carried out by employing theoretical–experimental methods explored through the bibliography.

Nonlinear optical processes are based on the expression of the properties of dielectric materials to establish a polarization that depends in a nonlinear way on the amplitude of the electric field through an intense electromagnetic field. It has also been stated that polarization is of utmost importance for the description of nonlinear optical phenomena. On the other hand, the nature of the atoms or molecules is of utmost importance in nonlinear processes due to their nonlinear properties at the macroscopic and nanoscopic level, as well as the type of bond and orientation of the crystalline structure. Thus, the physical origin of the optical nonlinearities depends on these properties. Besides, nonlinear optical processes stemming from the contribution of order-dependent nonlinear polarization were analyzed.

On the other hand, semi-organic crystals can combine the most outstanding characteristics of organic and inorganic materials in such a way that studies based on crystals as nonlinear optical materials are of importance for their potential applications in the field of laser technology and quantum data storage. An example of this is the use of crystals for the operation of high-energy laser systems. Some of the most commonly used crystals are potassium deuterium phosphate, lithium niobate, and barium titanate. The crystals have been used for the development of applications involving laser frequency doubles, electro-optical modulators operating in the infrared range, and phase conjugation. Furthermore, the saturable absorption exhibited by lithium niobate crystals can be used for ultrafast optical modulation in laser oscillators to generate ultrashort laser pulses. In addition, the nonlinearities exhibited by single crystals of potassium thiourea chloride are attractive for applications related to night-vision sensors, optical limiters, and photonic devices. Similarly, high-speed data processing is essential for many applications in the field of computing and telecommunications. In addition, third-order nonlinear optical response is critical for switching-related quantum computing applications. Optical architectures are highly beneficial for high-bit rate parallel and serial data stream processing systems. It is important to emphasize that the efficiency of optical data processing processes is highly dependent on the material used for the coupling of electrical and optical signals.

Organic materials and inorganic semiconductors are also good candidates for the development of nonlinear optical systems applications. Moreover, organic materials are of great interest due to their low cost and ease of integration in device manufacturing. Furthermore, the plasmonic behavior derived from decorations with metallic nanoparticles can function as an oscillator for frequency modulation. Furthermore, gold nanoflowers have a high conductive capacity due to their size and morphology. Because of this, potential materials are considered in the design and application of energy storage devices. On the other hand, metallic nanostars are considered ideal for biological detection applications. This claim is because they have high dielectric sensitivity, as well as near-infrared resonances. This is what defines nanostars as ideal structures for the development of substrates sensitive to localized surface plasmon resonances. Also, organic molecules and molecular materials are known to exhibit large nonlinearities and can optimize their properties by rationally modifying their structures.

7.5 Perspectives

The exploration of photonics has created a need for high-performance nonlinear optical materials with characteristics highly sensitive to applied fields. However, it is also necessary that nonlinear nanostructures have characteristics of optical absorption, processability, and mechanical, thermal, and environmental stability.

Thus, the field of nonlinear optics has also been widely explored to develop new biological research tools. In addition, nonlinear optics can serve in the determination of specific chemical strategies and high-speed imaging. Such strategies can be used for dynamic biomedical processes.

It was noted that there is a wide variety of optical linearities, and they are generally available. On the other hand, it has been observed that improvements in some classes of materials are necessary for the development of feasible applications, such that there are still endless opportunities to study materials.

It has also been highlighted that for the harmonic conversion and the optical parametric conversion of high-power lasers based on single crystals, inorganic crystals are usually a remarkable and important option. Molecules with large nonlinearities can be established for the design of devices focused on the electro-optical modulation of diode laser light in the near-infrared. However, it is necessary to think about improving molecular nonlinearities while maintaining transparency for such applications. Equally, progress has been made in the optimization of nonlinearities of organic crystalline materials that have exhibited nonlinearities greater than those related to inorganic materials. So, it is necessary to establish new strategies related to the growth and processing of crystals. Furthermore, the nanostructures that compose organic thin films are excellent for developing photonic applications. However, it is necessary to be careful when considering the physicochemical characteristics for the care of the quality of the films, such as the orientation, uniformity, and optical quality, as well as stability over time. Moreover, for third-order applications, it has been suggested that organic materials can have advantages over inorganic ones due to the magnitude of nonlinearity as synthesis methods.

It is also known that a great deal of materials research is still much needed. It is important to bear in mind that new concepts are necessary for the development of new applications and improve the efficiency of photonic materials. Therefore, it is also necessary to explore new methods for the manufacture and characterization of nanostructures.

References

1. Huttunen, M. J., Herranen, O., Johansson, A., Jiang, H., Mudimela, P. R., Myllyperkiö, P., & Pettersson, M. (2013). Measurement of optical second-harmonic generation from an individual single-walled carbon nanotube. *New Journal of Physics, 15*(8), 083043.
2. Min, W., Freudiger, C. W., Lu, S., & Xie, X. S. (2011). Coherent nonlinear optical imaging: Beyond fluorescence microscopy. *Annual Review of Physical Chemistry, 62*, 507–530.
3. Chen, M. M., Xue, H. G., & Guo, S. P. (2018). Multinary metal chalcogenides with tetrahedral structures for second-order nonlinear optical, photocatalytic, and photovoltaic applications. *Coordination Chemistry Reviews, 368*, 115–133.
4. Qian, H., Li, S., Chen, C. F., Hsu, S. W., Bopp, S. E., Ma, Q., . . . Liu, Z. (2019). Large optical nonlinearity enabled by coupled metallic quantum wells. *Light: Science & Applications, 8*(1), 1–7.
5. Jiang, B., Hao, Z., Ji, Y., Hou, Y., Yi, R., Mao, D., & Zhao, J. (2020). High-efficiency second-order nonlinear processes in an optical microfibre assisted by few-layer GaSe. *Light: Science & Applications, 9*(1), 1–8.
6. Kato, N. (2019). Optical second harmonic generation microscopy: Application to the sensitive detection of cell membrane damage. *Biophysical Reviews, 11*(3), 399–408.
7. Tran, R. J., Sly, K. L., & Conboy, J. C. (2017). Applications of surface second harmonic generation in biological sensing. *Annual Review of Analytical Chemistry, 10*, 387–414.
8. Sarri, B., Canonge, R., Audier, X., Simon, E., Wojak, J., Caillol, F., & Rigneault, H. (2019). Fast stimulated Raman and second harmonic generation imaging for intraoperative gastro-intestinal cancer detection. *Scientific Reports, 9*(1), 1–10.
9. Natal, R. A., Vassallo, J., Paiva, G. R., Pelegati, V. B., Barbosa, G. O., Mendonça, G. R., & Sarian, L. O. (2018). Collagen analysis by second-harmonic generation microscopy predicts outcome of luminal breast cancer. *Tumor Biology, 40*(4), 1010428318770953.
10. Canioni, L., Bellec, M., Royon, A., Bousquet, B., & Cardinal, T. (2008). Three-dimensional optical data storage using third-harmonic generation in silver zinc phosphate glass. *Optics Letters, 33*(4), 360–362.
11. Hales, J. M., Matichak, J., Barlow, S., Ohira, S., Yesudas, K., Brédas, J. L., & Marder, S. R. (2010). Design of polymethine dyes with large third-order optical nonlinearities and loss figures of merit. *Science, 327*(5972), 1485–1488.
12. Khan, Z. R., Alshammari, A. S., Shkir, M., Ganesh, V., & AlFaify, S. (2020). Enhancement in the photoluminescence, linear and third order nonlinear optical properties of nanostructured Na-CdS thin films for optoelectronic applications. *Journal of Nanoparticle Research, 22*(4), 1–16.
13. Condorelli, M., Scardaci, V., Pulvirenti, M., D'Urso, L., Neri, F., Compagnini, G., & Fazio, E. (2021). Surface plasmon resonance dependent third-order optical nonlinearities of silver nanoplates. In *Photonics* (Vol. 8, No. 8, p. 299). Multidisciplinary Digital Publishing Institute.
14. Arivazhagan, T., Vinitha, G., & Rajesh, N. P. (2019). Growth and characterization of diphenylmethanol single crystal by vertical Bridgman technique for second and third order nonlinear optical applications. *Journal of Crystal Growth, 512*, 181–188.
15. Naseema, K., Manjunatha, K. B., Sujith, K. V., Umesh, G., Kalluraya, B., & Rao, V. (2012). Third order optical nonlinearity and optical limiting studies of propane hydrazides. *Optical Materials, 34*(11), 1751–1757.
16. Fuentes-Hernandez, C., Ramos-Ortiz, G., Tseng, S. Y., Gaj, M. P., & Kippelen, B. (2009). Third-harmonic generation and its applications in optical image processing. *Journal of Materials Chemistry, 19*(40), 7394–7401.
17. Weigelin, B., Bakker, G. J., & Friedl, P. (2016). Third harmonic generation microscopy of cells and tissue organization. *Journal of Cell Science, 129*(2), 245–255.
18. Reyna, A. S., & de Araújo, C. B. (2017). High-order optical nonlinearities in plasmonic nanocomposites—A review. *Advances in Optics and Photonics, 9*(4), 720–774.

19. Davis, K. M., Miura, K., Sugimoto, N., & Hirao, K. (1996). Writing waveguides in glass with a femtosecond laser. *Optics Letters, 21*(21), 1729–1731.
20. Osellame, R., Hoekstra, H. J., Cerullo, G., & Pollnau, M. (2011). Femtosecond laser microstructuring: An enabling tool for optofluidic lab-on-chips. *Laser & Photonics Reviews, 5*(3), 442–463.
21. Hnatovsky, C., Taylor, R. S., Simova, E., Rajeev, P. P., Rayner, D. M., Bhardwaj, V. R., & Corkum, P. B. (2006). Fabrication of microchannels in glass using focused femtosecond laser radiation and selective chemical etching. *Applied Physics A, 84*(1), 47–61.
22. Galli, M., Wanie, V., Lopes, D. P., Månsson, E. P., Trabattoni, A., Colaizzi, L., & Calegari, F. (2019). Generation of deep ultraviolet sub-2-fs pulses. *Optics Letters, 44*(6), 1308–1311.
23. Midorikawa, K. (2011). High-order harmonic generation and attosecond science. *Japanese Journal of Applied Physics, 50*(9R), 090001.
24. Fareed, M. A., Strelkov, V. V., Thiré, N., Mondal, S., Schmidt, B. E., Légaré, F., & Ozaki, T. (2017). High-order harmonic generation from the dressed autoionizing states. *Nature Communications, 8*(1), 1–5.

Index

© The Editor(s) (if applicable) and The Author(s), under exclusive license to Springer
Nature Switzerland AG 2022
C. Torres-Torres, G. García-Beltrán, *Optical Nonlinearities in Nanostructured Systems*,
Springer Tracts in Modern Physics 287, https://doi.org/10.1007/978-3-031-10824-2

Printed by Printforce, the Netherlands